Walking The Path of
The Enlightened Jewish Master

A Unique Guidebook to Personal Mastery

This book is an adaptation of the original book:

Mastering Life:

A Unique Guidebook to Jewish Enlightenment.

published by Mosaica Press

Enlightened Jewish Master

The Enlightened Jewish Master

©Dov Ber Cohen 2024

All rights reserved. No part of this book may be used or reproduced or transmitted in any form or by any means, electronic or mechanical, without written permission from the author.

In memory of:
Shraga Feivel Ben David Binyamin
Riva Bat Moshe Tzvi
Michael Ben Meir
David Ben Arieh HaLevi
Dina Bat Chaim Getzil
Yitzchak Ben Kolonomus Kalman HaCohen
Mashkat Bat Tzvi HaLevi

and for the health and success of our children
Shira Mashkat
Chaim Michael Zachariya
David Binyamin
Riva Eliana

The Abramoff Family

In honor of the Holy Rav Dov Ber!
Mazal Tov on this amazing accomplishment.
May you continue to inspire the souls of Yidden around the world.

From Anonymous and Our Children
Yaakov, Bracha Dalia, Naomi Liebe, Levi and YY

Rav Dov Ber.
Thank you for all you do for Klal Yisrael.

Dedicated by the Pines Family.

Contents

Prologue 1

Introduction 2

1. **In the Beginning**: Choosing the Path of the Enlightened Master 4
2. **Sri Lanka**: Chilling with the Elephant Man 11
3. **Thailand**: The Philosopher and the Monk 24
4. **Laos**: Being the Go-between 39
5. **Korea**: Black Belt on a Paradise Island 52
6. **India**: Fasting, the Dalai Lama and Yoga by the Ganges.. 69

Photos 86

7. **China**: Shaolin Masters and Survival Missions 90
8. **Nepal**: Himalayas at Last 101
9. **Japan**: 40-day Pilgrimage and Aikido Training 116
10. **Israel**: Coming home and Walking the Land 127
11. **Why Judaism?** 142
12. **Soul Mates** 150
13. **Back to the East** 159
14. **The Continuing Journey to Jewish Enlightenment** 170
15. **Epilogue** 177

Appendix Contents

Practical Exercises to integrate the wisdom from each chapter.

A: Chapter One: Gratitude …………………………………..……178

B: Chapter Two: Mental Strength and Emotional Balance …………180

C: Chapter Three: Improving Character Traits ………………....……182

D: Chapter Four: Self Esteem and Being a Giver ……………..…..…184

E: Chapter Five: Self Control ……………………....……..………..186

F: Chapter Six: Discipline ……………………………………...……..188

G: Chapter Seven: Mindfulness ……………………………..…………190

H: Chapter Eight: Personal Accounting …………………...…………192

I: Who Wrote the Torah? ……………………………………..…………194

About the author, follow and contact …………………..………….208

Enlightened Jewish Master

In the merit of my father Joe Cohen
Unconditionally loving and supportive,
whose main teaching to us all was
"Just always do the right thing."

Doing my best to live up to that Dad.

Acknowledgements

First and foremost, my Mum and Dad - truly the most loving, supportive, understanding, sweet parents anyone could ever ask for. Everything I have ever done and who I have become is founded on the loving, healthy environment I was brought up in. Mum, you should merit to have a long, happy and healthy life in this world and the next and get great pride and joy from your children and Grandchildren.

All the mentors and guides I've had along the way.
The donors who supported this vision and made it viable.
My sweet, holy, wonderful wife, friend, soul mate. Couldn't do anything without you.

The Divine, Conscious, Loving Source of Everything - for creating everything for our benefit and guiding us along the way.

Life is a great challenge.
This battle is not external; it is raging deep within each of us.
It's to overcome our lower nature and manifest our Divine selves in the world,
to turn darkness into light, confusion into clarity,
challenge into growth, mediocrity to greatness.

It's a struggle to do the right thing in the face of adversity and pressure,
to live with the right values, to perfect ourselves and the world around us.
This battle is not fought with destructive weapons of war which harm
ourselves, others and the world around us; it is fought with mindfulness and
wisdom, love, consciousness and joy.

The way is one of learning to find peace, clarity, direction and connection;
to calm the mind, develop our character traits and manifest our true
spiritual selves against all the odds.

We are being lovingly watched over and directed by the One True Master;
The One who gave us the Guidebook to show us the way.
This battle is not to be feared or escaped; in fact
it's exactly what we've been created for.

Sometimes we'll fall, sometimes we'll get knocked down;
but with passion and perseverance we will surely prevail.
It's the path to True Mastery.

This, my friends, is what it means, to walk the path of the
Enlightened Jewish Master.

Introduction

"*I'm with the long haired yogi and the blankets, no sign of the donkey!*"

I heard a subtle tone of anxiety edging its way out through my wife's normally soft, calm voice.

"*That's fine,*" I replied in the most reassuring voice I could muster Possibly not the best response to ease that anxiety, but I had no time for formalities now as we were scheduled to set off early on our expedition, with fifteen Israeli backpackers, to volunteer in a village school in the foothills of the Indian Himalayas before the oppressive midmorning heat kicked in.

Granted, it wasn't your average conversation for a recently married, newly observant Jewish couple living in the heart of Jerusalem, and certainly not one I ever thought I'd be having with my wife after supposedly 'settling down.' However, just like so many other occasions throughout the years, I felt incredibly alive and comfortable with the situation, as the rumble of adventure began surging its way through my life once again. . .

You may have picked up this book because it sounds like an interesting story. While that may be so, the idea behind this book is much more than sharing what I've experienced on my journey. It's simply not enough to be just 'interesting.' The point of this book is to share the lessons I've learned, be impactful, to stimulate a shift and to inspire transformation. The point of this book is to encourage people to take an active role in writing their own life story in the most positive and meaningful way possible.

We all have a story. We all have lessons we've learnt, struggles we've overcome, experiences that change us, insights that give us new perspectives. We're all a complex combination of thoughts, feelings, conditioning, beliefs, hopes, issues, dreams and challenges. We're all writing and illustrating the story of our lives, moment by moment. One day we'll look back and, hopefully, enjoy reading what we wrote.

Know this; we don't just have to go along with things, hanging in there, trying to make the most of our circumstances, caught up in the challenges of life - surviving rather than thriving. We can live a life of empowered wonder, depth, meaning, adventure, growth and contribution. My goal is to help make sure we don't let our circumstances define us or how we feel and what we do. Allow yourself to dream again, to live fully and express yourself, to love yourself for no reason. Clarify the purpose and your goals and then work towards living them - joyfully. Let's live extraordinary lives, undertaking this magical journey together with the powerful words of Hillel the Sage in mind;

> *"If I am not for me, who is for me?*
> *And if I am only for myself, what am I?*
> *And if not now, when?"*[1]

---

[1] A leading Rabbi born around 50 bce, famous for saying "That which is hateful to you, don't do to others."

Chapter One
In the Beginning
Choosing The Path of The Enlightened Master

It was late afternoon. The shadows were beginning to lengthen across the lawn and the evening chorus of crickets started its nightly concert. The delicate purple flowers on the jacaranda trees were melting into silhouettes against the night sky and the smell of dinner began wafting through the hallways.

I was sitting with my grandfather in the garden of the nursing home in which he lived until his passing at 100 years of age. Although he was confined to a wheelchair for the last few years of his life, his mind was sharp right up until the end. My grandmother had passed away a few years earlier and although he had friends and family, he was left for much of the day on his own, reading books or lost in his thoughts. Breaking the silence of contemplation, he said something exceptionally powerful to me, something which shaped me and how I live, and will live, the rest of my life:

"You know Bradley,[2] you don't want to live an 'I should have' life."

After my prompt for clarification, he continued;

"You don't want to live a life that when you get to my age you look back and think to yourself 'I should have travelled more, I should have told my wife more often how much I love her, I should have spent more time

---

[2] My English name

with friends, I should have worried less what people think of me, I shouldn't have worked so hard, I should have asked if there is a greater purpose to life.' You can make sure you live your life now in such a way that when you get to my age you can look back in satisfaction at all the things you accomplished and experienced."

Insightful words from a man who had had a good, yet by no means easy, life and who now had time to reflect on a journey which was about to come to an end. At that moment, I fully accepted the challenge that he laid down and I committed to take responsibility for living the greatest life possible and ending it with no regrets.

Dinner arrived and it was time for me to leave. I gave him a kiss on the head and said

"Thank you Grampa, I'll do my best."

The foundation of piety and the root of all pure spiritual work is to clarify and verify what we are meant to be doing in this world.
- Path of the Just[3]

My sister and I grew up in a thriving Reform Jewish community in North London. We went to shul most Friday nights until my Barmitzvah, had Pesach Seders, lit Chanukah candles, I played football in the Jewish Sunday league, had Friday night dinner as a family every week and went to shul for the High Holy Days. I even visited Israel a few times - including living on a kibbutz for three months when I was 18 years old.

---

[3] A very famous book written by Rabbi Moshe Chaim Luzzato (Italy 1707 - 1746) about improving character traits and achieving self mastery.

In this way, although I never considered Judaism a viable spiritual path, I was imbued with a strong enough Jewish identity that throughout my travels, even though I ate snake, lobster, buffalo, live fish and a fine, 'tender and spicy' dog soup after my Taekwondo black belt test in Korea, I have never knowingly eaten pig; that would be going just a little too far for a nice North London Jewish boy.

I had a wonderful childhood, thanks mainly to my parents. The most loving, supportive and well-balanced parents any child could hope for, they were at almost every football (soccer), rugby, table tennis and tennis match I ever played in come rain or shine. In addition, they schlepped my sister and I to almost every art gallery and museum in the whole of Europe and ended up visiting me in Sri Lanka, Thailand, Korea, India and Japan.

It's very easy to live the adventurous life I did when one has a stable, supportive and loving base to go back to. I remember living in villages with no electricity or running water other than the river (two kilometers away) thinking that it was the best life, free of stress and responsibilities - until I realized that it only seemed so good to me because I could just get up and leave whenever I wanted, unlike the locals who were stuck there for the rest of their lives.

I really enjoyed school, managing to surf the border line between 'inspiring' the class and being kicked out. Although I often fell on the wrong side of the line, I generally ended up safely where I was meant to be due to the fact that my grades stayed consistently at the higher end of the scale. Towards the end of school, I started losing my way a bit as my natural teenage spirit for adventure and experiment began to take over. On Friday nights after Shabbat dinner I'd make my way to 'The Gallery,' a club near Kings Cross Station, and emerge into the sunlight eight hours later having danced the night away to some of England's top DJ's. Over the four years that I went to the five day Glastonbury music festival (the biggest of its type in the world), I saw some of the biggest names in music including Lenny Kravitz, Cyprus Hill, Bob Dylan, Chemical Brothers, Roni Size, REM, The Prodigy – as well as a group of pretty impressive Hare Krishnas who sang the same mantra more and more frantically non-stop for four hours inducing a trance-like state in everyone in the tent. Funnily enough, that came in handy many years later when I ended up in their headquarters in Vrindavan, three hours South of New Delhi and was able to join in with all their chants. I even dabbled a bit in dance music DJing, along with all the consciousness altering things that come along with that lifestyle. I played in some clubs in Manchester, but my 'highest' moment came at the Cambridge University end of semester ball, where I played an uplifting Drum n' Bass and Happy House set from 4.30 - 6.30 in the morning. Don't be fooled, given the right music (and substances) those academics in Cambridge really know how to let loose after an intense year of studying.

Despite all this, being good Jewish private school boys at heart, my friends and I never let the music scene take over our lives. It remained a passing phase from which we all emerged (mostly) unscathed. Although I spent more time in the gym than I did studying for my A-level exams,[4] I managed to come out with respectable enough grades to get into Manchester University to study philosophy - the only thing I could think of doing at the time. That's where things started falling apart.

University was a really bad time for me. It seemed to me that most people, including myself, weren't particularly happy or empowered. We were just trying to ride the waves of the challenges life was throwing us – trying to build a healthy sense of self amongst difficult relationships, unfulfilling jobs, messed up values systems, stressful exams and an overall sense of meaninglessness. I started an existential experiment. I walked around asking people 'How are you?' The general response was something along the lines of 'Can't complain,' 'Getting by,' 'Not too

---

[4] State exams at 18 years old

bad' 'Fine' 'hanging in there' 'could be worse' and other such lackadaisical answers. It seemed like a tragedy to me that this was the general feeling in the world I lived in. Looking towards my future, it seemed like most people weren't loving their jobs, there was a high divorce rate even then, depression and anxiety were building and there was a general escape into drug use and entertainment as a way to get a break from life for a few hours.

Even looking at the many people who were actually living decent lives, I noticed that it wasn't because they had sat down to really question what the purpose of life actually is and then commit to working towards that. They were just going with the flow, not asking too many questions, making the most of things and life was going well for them. I realized that most people just become a product of their society and upbringing, without ever really stopping to think what they believe, why they believe that and how they know that what they believe is actually true. I started to wonder whether there is a greater point to it all, something beyond the surface that we are meant to be tapping into and achieving. I understood, or rather hoped, that daily life in the West can't be the be all and end all. There has to be more to life. I had a choice to fight for that goal or to go along with the flow, and that flow didn't look like it ended up anywhere particularly appealing.

I wanted to thrive, not just survive - as William Wallace says in Brave Heart, "Every man dies, not every man really lives."

I felt like I wasn't really living. At that point, having reached the low point of my life, I was either going to crash or dig really deep and reach really high. Like most things in life, the greater the stretch, the greater the potential for growth; something I experienced as I was catapulted back up from the end of the 216 meter drop having just jumped off the bridge of the world's highest bungee jump, another crazy attempt to seek out life's peak experiences.

In my final year of university, I stopped all unhealthy habits, didn't miss one lecture, started going to the gym, meditating, doing tai chi, karate and going for long walks in the park to find space and clarity. I read classic books about personal journeys including Siddhartha by Herman Hesse, The Peaceful Warrior by Dan Millman, The Alchemist by Paulo Coelho and The Celestine Prophecy by James Redfield and started really contemplating life. I suddenly felt totally alive, searching for the truth, for the diamonds buried deep down somewhere just out of my reach. A note scribbled on the side of one of my lecture pads from that time reads - "My mind is getting too big for my head." It was a different sort of mind expanding experience than the ones I was used to. Like a butterfly emerging from a cocoon, I was finally breaking out of my old thought patterns and conditioning and starting to take control of my destiny.

I began spending hours in the library reading about mysticism, Eastern philosophy and meditation. The teachings of Buddhism resonated with me. Their ideas focused on detachment from the physical world (which causes suffering), controlling the mind which plagues us constantly, correcting negative character traits and generally helping us rise above life's hardships and challenges. In their view, it is impossible to be involved in this world, with its materialism, physical pleasures and emotional trials without it leading to suffering. Their answer is to retreat from engagement in the physical world, seeking spirituality in natural caves and secluded temples. It sounded like just what I needed.

I started watching martial arts films and saw people learning to have discipline, train their body and mind, rising beyond the normal experience of life. It was then that I started planning my 'Warrior's Journey.' I decided that after leaving university I would travel the world, discover the purpose and meaning of life and then create something based on what I learnt. I was set on living my dreams, experiencing life and the world fully rather than just finding a job straight away without knowing clearly what life is about and what I was living for.

In my mind at that time, all true wisdom lay to the East; in the peaks of the Himalayas, colorful temples in the jungles of Thailand and at the feet of old, white-bearded masters in the mountains of Japan. There I could find real sages, learn to control the mind and body, rise above the sufferings of life and reach true mastery and enlightenment. It didn't even cross my mind to consider Judaism, which to me was at best just a nice cultural tradition and at worst was an outdated, male chauvinistic, naive, immoral set of rituals followed by a group of people who didn't even seem very nice, let alone enlightened.

Of course, at that time I'd never come across the likes of real Jewish masters, from Rebbe Shimon Bar Yochai, The Arizal, the Ramchal, Ba'al Shem Tov, The Vilna Gaon, Rebbe Nachman, the Chofetz Chaim, Rav Kook, the Lubavitcher Rebbe and Rav Moshe Feinstein to name but a few – Jews who throughout the generations were blazing a path to ultimate self-realization and G-d consciousness based on ancient wisdom, mindful living, authentic prayer and deep meditation.

By the end of university, I was ready to start the real adventure of life, to set off to the Lands of the Rising Sun, in search of true meaning and enlightenment. The journey of the Enlightened Jewish Master in training was about to begin.

Chapter Two
Sri Lanka
Chilling With The Elephant Man

Sri Lanka, situated just thirty-one kilometers off the southeast coast of India, is to this day one of the most vibrant, lush, tropical, warm, friendly and culturally diverse places I've ever been. Oozing with ancient culture, scattered throughout the island one comes across hidden caves displaying exquisite wall art dating back to prehistoric times, ruins of once thriving civilizations, temples lined with ornate sculptures and rock fortresses sprawling up into the forest covered mountains. The landscape is breath-taking; boasting world famous tea plantations stretching out as far as the eye can see, mile upon mile of thick rain forest which plays home to thousands of species of tropical flora and fauna and some of the best palm-lined surfing beaches in South East Asia.

This paradise island was the first stop on my journey of self-discovery and a far cry from the 'developed' world that I had left behind. Sri Lanka was the most random place I could think of at the time; yet from the moment I arrived, I felt more at home and alive than I ever had before...

After receiving my BA honors degree in Philosophy and completing a month long intensive TEFL (Teaching English as a Foreign Language) course, I booked my one-way ticket and set off on the first leg of my epic adventure in the East. After a two-hour journey from the airport I arrived in Kegalle, a small yet bustling town halfway between Colombo the capital and Kandy (the cultural center of Sri Lanka). I had never seen so many people in one place at the same time, each one going about their business, a cacophony of sights, smells and sounds, bus and motorbike horns, street vendors calling out their wares, all interlaced with the barking of dogs, mooing of cows and bleating of goats, which were all roaming freely around the busy streets. Although the monsoon season had officially ended the rains were still in full force and muddy rivers overflowed their banks, flooding the already poorly paved roads. The buses were crowded to bursting point, with people literally hanging out the doors and sitting on the roof. Yet, despite this, I never saw anyone lose their temper or composure. I guess that when you are poor there is less sense of self, less haughtiness and sense of 'me and mine,' of 'do you know who I am?-ness.' Everyone is in the same creaky boat trying to make it to shore together.

In general, Sri Lankan people are humble, unassuming and friendly, getting on with their day to day life, trying to eke out a living and put food on the table, devoutly stopping in to the local temple to pray each morning and evening. I spent a couple of weeks in Kegalle teaching English to the staff of an agricultural co-op called Sanasa and working in the local orphanage. Although there were few other westerners there, I felt like getting away from everything I knew and my main pastime when I wasn't teaching was building a hut in the forest for the rubber tapping workers to rest in in the midday heat. My tender, middle class hands which had never seen a real day's work were soon blistered and bleeding from the hoeing and chopping, but that didn't deter me. I ripped the sleeves off my t-shirt, tied them round my hands and continued hacking away, much to the amusement of the locals who, to be fair, were actually doing most of the work. I paid the price for that act of bravado for the next week or so, as it made eating a meal of hot rice and curry with my bare hands that much more painful; another good source of amusement for my new friends.

Down the road from the center was an elephant orphanage where I used to go to help out and sometimes ride the elephants which on command would take up a trunkful of water and spray it backwards onto their unsuspecting rider. I soon became good friends with one of the workers there, Senerat, who was around ten years my senior and had come from northern Sri Lanka. He was trying to make a go of his life in the central region which had been less affected by the vicious and drawn out civil war. He had learnt English and become pretty westernized, touchingly proud of his denim jacket that he'd wear every day whatever the weather. We used to climb up into the forests to temple ruins where we'd sit drinking beer and comparing cultures - he'd teach me Sinhala and I helped him with his English slang. Behind his large, soft, brown eyes something betrayed his difficult past, and one drunken night he shared his ordeal with me. Five years earlier he'd seen his village in northern Sri Lanka overrun by the LTTE[5] who massacred much of the population including most of his family. Apart from Senerat, just his brother, sister-in-law and their three kids managed to escape. They were now trying to rebuild their lives somewhere near a lake in the middle of the country, living from hand to mouth as do most Sri Lankans.

I was made acutely aware of what a cushioned and privileged life I had been blessed with, far removed from the suffering and horrors so prevalent throughout the world. Some people feel very guilty about their privileged upbringing, asking why did I deserve this, why is this fair? For me, guilt didn't seem like an appropriate feeling, seeing as I had done nothing wrong. What I did feel, though, was a tremendous

---

[5] Liberation Tigers of Tamil Eelam, a militant group fighting for independence in Northern Sri Lanka

responsibility to share my privileges with those who weren't lucky enough to be born with them.

Very soon after my arrival in Sri Lanka, and possibly for the first time in my life, I really felt that life is not all about me and my needs, my relationships, my dreams, my money. Life is much more fully lived when the focus is not on myself, but on how I can be there to improve the lives of others. How do you really become a bigger person? It came clear that the key to becoming a bigger person was to learn to step beyond my 'Self' and incorporate the needs of others into my life goals. Someone living with values and ideals, who is willing to sacrifice some of their ego-needs for the greater good, is someone who is truly living a fulfilling and meaningful life.

"The more one has, the more one wants"
Sri Lankan Proverb

"Who is rich? The one who is happy with his lot."
Ethics of Our Fathers 4:1[6]

I left England with just nine kilograms on my back: A couple of pairs of pants, some shorts, some t-shirts, underwear and a sweater and even that was more clothes than most of the people I was to meet on my journey had. Added to that, I had a mosquito net, a ten-meter piece of string, a head torch, writing pad, padlock, a diary, mosquito spray, swiss army knife, toothbrush and paste, painkillers, antiseptic cream and a couple of books.

---

[6] A 2000 year old Jewish text of short pieces of advice from our Sages.

There's a saying in India 'The poor man sleeps well.' He has nothing to lose. He doesn't have to worry about his car getting scratched or his stock going down. Similarly, living out of a backpack leaves you little choice: you can only carry what you really need, and it is incredibly liberating to realize just how little one needs to get by. Having limited 'stuff' taught me to really value, take care of and appreciate everything I had, as well as learn a lot about myself and what is important to me. My ten-meter piece of string was essential for putting up the mosquito net, hanging washing up to dry, building shelters and repairing my backpack. I valued it immensely and rolled it up and packed it away carefully each time I moved on. It's amazing how something so cheap and easy to replace can be so valuable! In Asia, insect repellent is indispensable if you want to have any peace of mind, whether walking through a jungle, meditating on a mountain or sitting around a fire on the beach. I guess I'd add on to the Indian saying - "the poor man sleeps well ... as long as he has a mosquito net."

However, for me, having very little was a choice, unlike for the people who I was spending the bulk of my time with. From working in the orphanage in Kegalle, then time and again from the attitude of people who were surviving on the bare minimum, one meal a day and one set of clothing, in Sri Lanka, India, Malawi and Thailand, I really learnt one of the essential lessons in life and main keys to happiness; appreciation. For these people life was certainly a struggle; but that's what made them grateful for everything little thing they did have and they seemed genuinely more content than many of my friends back home. I once bought a pencil for each of the kids in the orphanage, who until then shared one pencil between five of them. It was the most moving sight to see the joy in their eyes, the authentic wonder at having their own pencil. Every meal they received, every smile or hug or game we shared brought so much delight to their lives and taught me that it is not what you have but rather how much you appreciate it that creates true happiness.

It seemed to me that so often I'd focus on what I didn't have, saying to myself "If only I had a better job/car/phone/computer/relationship/more money, looked different ... then I'd be happy," or "when I'm rich/married/have kids/have more free time/get healthy ... then I'll be happy." It became so obvious that focusing on what we don't have or comparing ourselves to other people who do have those things only brings us down. I decided to start focusing on and fully appreciating what I did have, and quickly saw the positive effect it was having in my life.

I soon noticed that apart from more contentment in life, this attitude has many other benefits. Not only does it breed happiness, it also helps get rid of the burden of such emotions as envy, jealousy and greed, seeing as once we appreciate what we have, what does it matter what others have? At that point we then become more willing to help others and celebrate their joy with them, rather than putting them down or focusing on taking for ourselves in order to fill in our sense of lack. In fact, a different understanding of the verse quoted above from Ethics of Our Fathers teaches us just this.

Who is rich? One who is happy with his lot' can also be read as Who is rich? One who is happy with '*his*,' meaning the other person's, lot. It's a high level to be truly happy for someone else.

On arrival in Israel almost seven years later, I learnt that in many ways gratitude is the essence of what it means to be Jewish. For a start, the word for Jew in Hebrew is Yehudi which comes from the word 'lehodot' meaning to "thank, acknowledge, or admit". In other words, our very essence is to be grateful, count our blessings and appreciate what we have. It was with this in mind that our sages taught us to begin our day with our first conscious thought as Modeh Ani – Grateful am I.[7] Seeing as so much of the potential for something comes at the beginning, to begin the day with a (conscious) declaration of gratitude is really the best way to make sure we avoid 'getting out of bed on the wrong side.'

There is a famous story of two men who went to visit Rabbi Zusya of Hanipoli, famed to be a great Tzaddik (righteous person) and master. He was extremely poor, living in a rundown shack, surviving on the bare minimum often without enough food to feed his family. They arrived at his door and were shocked that the stories were true. His shelves were empty and he was wearing worn out rags. They asked him "How can you say the daily blessing 'Blessed are you Lord our G-d who has given me everything I need?'" Rav Zusya calmly looked around, broke into a wide grin and replied "This must be all I need!"

From this, we see that one of the foundational roles and attitudes of humans in this world is to recognize the good in our lives. This doesn't just make us happier people. On a higher level it is this attitude which forms the basis of what our sages teach is the purpose of creation; to build a deep connection to the Source of Good in our lives. By appreciating all the positive things in our life and recognizing where they come from we can create an authentic relationship of awe and love with the Creator.

---

[7] In Hebrew grammar it should really be Ani modeh - I am grateful, but our Sages didn't want the first word in the morning to be 'I' - ego, rather we must start the day with gratitude.

Whenever I ask someone what they want, they invariably say 'I want to be happy.' Everyone does. Sadly, very few people are authentically and consistently happy. It's because they haven't defined what happiness is and how to achieve it. If we think happiness is the feeling I get when things go my way, we'll spend our lives fixated on trying to get the external world to meet up to our expectations and desires - an impossible task. Judaism teaches that real Simcha (joy) has nothing to do with the external world. Simcha is an attitude to life. a state of being. The Baal Shem Tov[8] teaches us the key to happiness. He points out that the word b'simcha (with joy - בשמחה) is made up of exactly the same letters as machshava – thought (מחשבה). If our thoughts are in the right order, ie we have a positive outlook on life, will have a stable state of true happiness, not reliant at all on external circumstances beyond our control. Our attitude defines our experience of reality. Someone with a positive attitude will deal with even the hardest challenges in life, and someone with a negative attitude will always find something to complain about even if they are one of the most privileged people in the world. With this in mind, one of the key attitude shifts[9] is for us to go from focusing on what we are lacking to always finding the best in things, focusing on and being grateful for every little blessing we have in our lives.

---

[8] (1698 - 1760) - the founder of the Chassidic movement within Judaism which stresses consciousness, joy and simplicity in our spiritual expression
[9] We'll be looking at several others over the course of this book

Enlightened Jewish Master — Sri Lanka

"I have written this not to teach people what they don't know, rather to remind them of what they already know very well. However, just as much as the ideas are well known, so too the neglect of them is common and forgetfulness is great."

Path of the Just

Everyone reading this book knows the advantage of living with gratitude, could probably teach a class on it and even tell a couple of cool stories to back up their point. But if after the class they get upset when they don't get the food they wanted, or they are jealous of their friend for buying a new car, the ideas meant nothing. As with everything in life (and all the ideas in this book), it's not just enough to know and understand them. Ideas alone are almost useless.

Real transformation only comes from daily practice, discipline and integration of the knowledge, turning it into wisdom. How many of us are actually taking the time to do the real work to implement and live with these ideas? How often do we read books like this and skip the exercise part or comment on what a good exercise that would be, but never get around to doing it? It's no good only appreciating our legs just when they hurt or when we see someone in a wheelchair. There has to be a conscious appreciation of our blessings several times throughout the day in order to ensure the promise of happiness this idea brings. That means not just recognizing and even listing what we are grateful for but taking time to actually *feel* a deep sense of gratitude.

For each chapter of this book there is an appendix with some suggested exercises and practices you can implement in order to integrate these ideas into your daily life. I strongly advise investing some time in them. See Appendix A for some tips for developing and experiencing gratitude.

Working in the orphanage in Kegalle woke me up very powerfully to this truth. The kids there appreciated their small blessings more than many of us appreciate our big ones. Seeing these kids and seeing how little it took to deeply affect their lives, helped me realize how much I have and therefore how much I have to give.

After two weeks in Kegalle, I felt that I needed to go somewhere totally on my own, even more remote, far from any civilization as I knew it. The option came up for me to go to the Sanasa branch in Monaragala, fifteen hours by bus south east of Kegalle. Monaragala was the sort of place that even Sri Lankans would wonder what on earth I was doing there. On one bus ride between Kandy and Badulla (the closest town to Monaragala) through the pouring rain, on a bus with maybe half its windows intact and no windscreen, on typically treacherous Sri Lankan roads, I was chatting to a teacher from Colombo. We were all squeezed as far into the middle of the bus as we could, trying hopelessly to avoid getting wet, praying that the fact that we were driving fast along dangerous mountain roads (with a driver who constantly had to wipe the water off his face) did not end in disaster. When I replied 'Monaragala,' the teacher who had asked me where I was going burst out laughing and said "Suddhu mali, (white brother) you are crazier than you look!"

I spent the days teaching English, washing my clothes on a rock and meditating in the local temple, where I would go each day for early morning and late afternoon prayers. I soon learnt all the words to the chants and could join in with the monks in their sunrise and sunset services. I became very friendly with Rohan, the Sanasa caretaker and his family and used to sit in the kitchen with them helping peel the potatoes and carrying the huge cauldron of boiling rice from the fire to the dining room, laughing and joking as we tried to communicate in broken English and Sinhala. I didn't really realize what a stir I was causing as I was smashing the class boundaries, ignorant of the caste system which put the white English teacher far above the kitchen workers who were his best friends and closest confidants. It was one of the best times in my life, very simple and pure, unlike anything I had ever experienced, totally on my own, learning and growing, meditating, exercising and reading, interacting with genuine and friendly people in the most random place in the world; I was truly in my element.

Towards the end of my time there, Senerat, the man from the elephant orphanage in Kegalle, decided to take all the English volunteers to his brother's village for a taste of even more rural Sri Lankan life. We met in Kandy and after a train journey, a bus ride sitting on the roof and an hour hike through the countryside, we arrived in the 'village' which actually turned out to be just one hut in the middle of a field. We were greeted warmly by the family, white teeth shining out through their wide and genuine smiles. The field was covered in small white, round-headed flowers, I think chamomile, which Senerat's six-year old niece ran out to collect. Five minutes later, the pot was on the boil over the fire and we all enjoyed a cup of some of the most interesting tea I've ever tasted. After talking to the family with Senerat acting as translator, we met up with an old man and a few younger guys from a nearby village and went for a hike up into the mountains which surround the lake on all sides. As we scrambled over the final rocks to reach the plateau on top, we were met by a truly mind-quietening sight. The deep orange sun was setting behind the vast plains and I was suddenly hit by a transcendent feeling of awe at the vastness and beauty of creation. It was a moment of epiphany as I was left with the realization that I am just a small part of a much greater whole, that there is so much more to life then I could ever have imagined. I became aware that there is so much beauty, so much to achieve, so much beyond me and my personal needs, so much beyond the grind of the 'real world' I had left behind - it made me feel small and humble, and yet at the same time more connected and empowered than I had ever felt before. We made it down after dark and found a suitable place to camp for the night. While the others were having a rest, the village guys invited me to go fishing with them in the lake. The sun had already set and there was no light pollution (or electricity at all) in the area to dull the brilliant glow of the myriad stars illuminating the scene around us. Just before entering the lake, they gave me my weapons: a dull flashlight and a large knife. As I waded through the murky water straining to try to catch sight of a fish I took the opportunity to 'zoom out' of my life to locate myself in the world, as if looking down from space, getting a new perspective on the present moment and the absolute absurdity of what I was doing. The current scene was particularly random: a nice Jewish boy from North London standing knee deep in the middle of a lake in central Sri Lanka, flash light and 'spear' at the ready, hoping to catch dinner. I didn't see a fish, let alone catch one, but the others seemed to be more successful and managed to provide the group with plenty to eat for dinner. (I sometimes wonder if they were just having fun with me, and really they had bought the fish in the market earlier that day). We got back to the campsite where Senerat's sister in law started preparing the fish curry and rice for dinner. While the others sat in a circle around the fire, chopping vegetables, drinking beer and sharing stories, Senerat said I should go with the old man to his village to collect some mats to sleep on. The man, with a strong smell of whiskey on his breath, had a rickety motorbike, which didn't look as though it could actually start up, let alone carry us both. As I got on the back of the bike, Senerat came over to me and whispered 'take this with you brother' as he pressed a big knife into my hand. Before I had time to protest I was whisked off on a crazy 50 km/h motorbike ride through the pitch black Sri Lankan night, no roads or even dirt paths to be seen, drunken old man for a driver, having just been secretly slipped a large knife, I assumed for protection should I need it. Twenty minutes later we somehow managed to arrive in a village of ten or so huts, and the man told me to wait on the bike

while he collected the mats. I sat there clutching the knife, looking around me, waiting to be attacked from all sides. It occurred to me that even if I did manage to defend myself somehow, I was totally lost in the middle of nowhere, prey to the elements and whatever dangers were lurking in the Sri Lankan wilderness. Ten tense minutes later, the old man returned with twelve straw mats and we made our way back to the campsite, 50 km/h again, but this time with the added hazard of trying to hang on to the mats on my lap with one hand and the driver with the other. When we got back the mouth-watering smell of freshly cooked fish curry, a large camp fire and some beers cooled off in the lake were waiting for us. Senerat came up to relieve me of the mats
'Um...Aya,' I addressed him like always as 'brother.'
'Yes brother?'
'Why did you give me the knife?'
'Oh no problem brother. Just, I've never met that man before and there are dangerous people in these villages, he could have tried to harm you.'
'Thanks Aya, very thoughtful of you,' I said, resting my hand on his shoulder. At least he had the courtesy to give me a knife when he sent me off into the wilderness on a rickety old motorbike with a drunk and possibly dangerous local villager.
After eating my fill, I lay down on my mat, looked up at the stars and feeling truly grateful that I was alive, well and well fed, as I fell asleep to the sweet sound of traditional Sri Lankan folk songs which pierced the starry night sky and echoed off the mountains all around.

Chapter Three
Thailand
The Philosopher and The Monk

*I*t's 4am. I'm roused from my uncomfortable sleep on the wooden bed with the wooden 'pillow,' in a room barely big enough for the bed and my bag, by the incessant ring of the bell. No, it's not Thai prison, it's Wat Suan Mokh ten-day silent meditation retreat in Surat Thani Province, Southern Thailand. I'm here with about fifty other foreigners all seeking peace of mind, mastery of thought, a quenching of desire, Nibana[10] – Enlightenment.

Emerging from under my mosquito net I make my way outside from where I hear the splashing, as the other still half asleep participants are already pouring buckets of cold water over themselves from the huge cement troughs in the courtyard. I join in the ritual: nothing like a cold bucket shower to wake you up in the morning. We trudge towards the meditation hall as the first greyish light begins to reveal the silhouettes of the palm trees lining the surrounding hills. Fifteen hours later, we fall back into our wooden beds, another day of conscious breathing, conscious walking, conscious eating, consciously coming face to face with ourselves in a very real way. Away from any distractions - no TV, no phone, no reading, no writing, no computers, only one meal a day - just us and our thoughts, we are learning to focus all our attention on

---

[10] In the Pali language, used in Buddha's time. More commonly known in the west by its Sanskrit word: Nirvana

our breath, trying to discover the inner peace, bliss and higher awareness that lies underneath the constant babbling of the monkey mind.[11]

From Sri Lanka I went to Thailand and moved straight into my new home; the small village of Nong Pau Tan. Situated in the middle of paddy and pineapple fields 200 km south of Bangkok, Nong Pau Tan lies in the thinnest part of the country, nestled between the mountainous Kan Krachaen National Park and the pristine coast of the Gulf of Thailand. The old wooden houses, badly paved roads and population of simple farmers belied the large TVs and computer game consoles that could be found in many of the houses – it became clear that some form of digital entertainment to unwind after a hard day's work, whether in the office or the paddy fields, is a top priority throughout the world, no matter what the situation and standard of living.

My host was Kru Salee, a forty-five year old school teacher who had been a student of mine in my English teacher training course in London. We hit it off immediately and she invited me to stay with her family and come to teach in the village school. As my time in Sri Lanka was drawing to an end, I took her up on the offer and so started my one-and-a-half year sojourn in the Land of Smiles.[12]

The first couple of weeks, as was the case in most of the countries I lived in, were pretty lonely and challenging. Being the only farang (foreigner), looking so different from everyone else, unable to communicate or understand what anyone was saying about me was very disempowering and humbling. Salee introduced me to the boys in the village and once again, as has happened so many times throughout my travels, the saving grace came in the form of the good ol' universal

---

[11] A term referring to the nature of the mind which is always unsettled and jumping around, often causing trouble
[12] Popular nickname for Thailand

language – football.[13] Bring a group of boys together, give them a ball and they'll be best friends, or worst enemies, within half an hour.

Every day after school and work, all the guys in the village would gather at the football pitch, pick teams, and slog it out for the next two hours. After the game we would go back to Gurn's house, get a bottle of cheap Thai whiskey and some cold beers and tuck into some local delicacies; Som Tam – spicy papaya salad, kuay tiow – noodle soup, and once in a while someone would grab a duck which was innocently waddling around, twist its neck and throw it into the bushes until it stopped struggling. They'd pull off the feathers, boil it up with some spices and vegetables, and within half an hour everyone would be relishing a fine duck soup to go with the whiskey and beer. The host, Gurn, was about 4 foot 6, with gnarled little legs, six toes on each foot and six fingers on each hand. How he was quite such a good defender I'll never know, but there was hardly a time when both the man and the ball went past him, usually it was just one or the other. After volunteering in the village school for a few weeks, I got a job teaching English in a university about a forty-minute drive up the road towards Bangkok, and between my football buddies, university students and Kru Salee, I was speaking highly conversational Thai within three months.

Socially settled, financially stable and now able to communicate, it was time to go in search of the master and the practice that was the motivating factor for my travels in the first place. I found him in the form of Tan Medhi, a forty-year old Thai man who had renounced his life as a lawyer in Bangkok and was now living the ascetic life (replete with shaved head and orange robes) of a monk in the forest temple which became my second home in Thailand. I would join him on his early morning alms round to the nearby villages and we'd discuss life, meaning, meditation and philosophy for hours, before finding our own spot by the lake where we'd settle into our meditation practice as the

---

[13] Soccer

chirping of the cicadas and croaking of the frogs helped lull us into a peaceful state of mind. We always joked that this would make a good photo for the cover of our first book - The Philosopher and The Monk. After my initial efforts in my Sri Lankan temple, my real quest for mastering the "monkey mind" had truly begun.

"You are what you think. All that you are arises from your thoughts. With your thoughts you make your world."

Dharmapada[14]

"It is important that you do not ignore taking care of guarding your thoughts ... for the majority of imperfection and rectification of one's action is only through them."

Duties of the Heart[15]

Originating in the Chinese and Japanese Buddhist traditions, in the East the mind is often referred to as 'monkey mind' – constantly jumping around, rarely still and at rest. From the second we wake up to the second we fall asleep we have a constant companion; commenting, judging, comparing, doubting, desiring, worrying, regretting, and generally causing all sorts of mischief. The mind has taken over and rather than us controlling it and using it when we need, it is running wild, often to negative places. If you had a friend who spoke to you how you often speak to yourself, how long would you let them be your friend? Try to stop thinking for ten seconds and just focus on your

---

[14] Core text of Buddhist teaching dating from around 250 BCE
[15] Another famous book about character development and self perfection, written by Rabbi Bachya Ibn Pakuda in Spain in around 1080ce.

breath and you'll see it is almost impossible. How many times have you lain in bed wanting to sleep but you can't because your mind is in overdrive? How many times have you gone over the same chain of thought - whether it be rehearsing a speech you want to give to the person who hurt you (which you often never actually share), replaying the mistake you made or worrying about what other people at the party thought of you or if you remembered to turn off the oven or lock the door?

The mind is the backdrop to everything we feel and do, ultimately creating our whole experience of reality - for good or for bad. Most of us are controlled by external circumstances; if we miss the bus we get frustrated, if someone compliments us we feel great, if it's raining we're down, sunny - everything is good in the world. The external world acts as a stimulus which is filtered through our beliefs, creates thoughts and those thoughts create our feelings.

Let's do a thought experiment to demonstrate this.

You go to the bank and get in line, and someone pushes in front of you. How does it make you feel? How do you react?

The truth is that it depends on many factors, for example:

- What you believe and how you feel about yourself.

If you have low self-esteem you think 'I can't believe I let someone do that again,' or 'I deserved that' and you'd then feel bad about yourself and cry or withdraw to your room. Yet if you are arrogant[16] you may think "How dare they? Do they know who I am?" and you'd feel anger and shout or worse.

- It also depends on what you think of the other person; if you notice it's a close friend you haven't seen for a while you'll react very differently than you would if it is someone you don't know or don't like.

---

[16] Which is another form of low self esteem - needing recognition from others.

- It also depends on your mood; If you just had a big argument with your boss and stormed out, you may make a fuss, if you just won the lottery you may just let it go.

What this demonstrates is that it is not actually the external circumstances that are dictating our experience of reality, but rather any number of internal factors and beliefs.

In Judaism, our unhealthy drives and destructive mind patterns are referred to as the Yetzer Hara – the inclination towards negativity. It prevents us from moving towards achieving great things, and guides us towards doing, thinking, feeling and saying things which are harmful for us and those around us.

It is surprising, then, that our Sages teach that the yetzer hara is our best friend. How could this be? It's because only by overcoming, conquering, taming and ultimately redirecting its energy can we become true masters. This is expressed in what is (for me) one of the most essential teaching in Ethics of Our Fathers:

Ezeh Hu Gibor? Who is the strong one / master?

HaKovesh Et Yitzro - The one who conquers their own inclination towards negativity.

The real master is the one who has learnt to tame the monkey mind, to quieten the chatter and retain awareness on the positive. This is incredibly empowering, seeing as it means it is really up to us. We can't often control what happens, but if we are conscious enough we can certainly control how we let it affect us.

My daughter, Tamara Menucha, who was three at the time, learnt this lesson in a very powerful way. One Shabbat we were sitting together reading a book and she asked me for some ice-cream. I went to the freezer and was faced with a terrible situation – no ice-cream! I returned to Nucha, as she liked to be called, and informed her of the situation, to which she quite understandably reacted by starting to whine. Now, although it is important to learn to control our thoughts, feelings need to be dealt with in a different way; they need to be acknowledged and accepted. Sometimes when we feel upset about something, or jealous, we beat ourselves up saying 'I shouldn't feel upset, I'm bad for feeling jealous.' This doesn't help the situation at all - not only do we feel upset, we then make ourselves wrong for feeling that way, which makes us feel even worse! We need to learn to accept our feelings and just feel them, give them space to manifest and express in a healthy way, and in this way they will pass quicker. In fact, Rashi[17] comments that all difficult things in life are because

'our consciousness is not wide enough to accept them and we don't have room in our heart just to feel the pain.'[18]

Only once we do this, can we work on controlling the thoughts that could perpetuate the negative feeling.

So I tried to validate Nucha's feelings, telling her that I hear she's upset, it's quite understandable and that I was upset too because I really wanted ice-cream. Nucha, not appeased, continued to whine. Now time for the lesson:

Abba: 'Nucha, I've got a question. Do you think whining will make more ice-cream?' (Nucha looks confused).

Abba: 'Look, I'm getting a bowl from the cupboard and we'll whine together and maybe we can make ice-cream (Abba gets bowl and brings it to the table) …. Aarrrhh, noooo, it's not fair ….'

(Nucha smiles. From whining to confusion to laughter – good progress).

Abba sees his chance: 'Nucha, I want to teach you an important lesson. We both want ice-cream, but there is no ice-cream and we can't get any. So now we have a *choice*. We can *choose* to be upset by this for the rest of Shabbat or choose to be okay with it. If we want to be sad we can have these thoughts: 'It's not fair, I want ice-cream, I bet my friends have ice-cream, I'll never have ice-cream, I hate not having ice-cream

---

[17] Rabbi Shlomo Yitzchaki (France 1040 - 1105) The foremost Torah commentator
[18] Bamidbar 21:4'

...' and if we want to be happy then we should have these thoughts 'I can have some cake instead, I can play with my toys, I'll probably have ice-cream tomorrow, I'm happy for my friends that they have ice-cream....' So really, we can choose how we feel, it's up to us!'
Nucha: (after some thought): 'Okay Abba, let's choose to be happy."
(Nucha takes a cookie and we go back to reading the book).

This was an awesome lesson for a three-year old to understand, possibly the most important lesson she will ever learn. Whereas as we can't often choose our circumstances and control the world around us, we can start raising our consciousness and choosing how to relate to what goes on and therefore how we feel and what we do about it. [19]

All of us are going to face pain and challenges in our lives, both physical and emotional. Our job is to be conscious and disciplined enough to be able to acknowledge and experience our feelings in a healthy way, not reacting or repressing but rather processing what happened and moving on in a positive way. We can be a victim to our

---

[19] I'm not, G-d forbid, belittling anyone's suffering and suggesting they just change their perspective so easily. It is hard enough to control the mind and choose to be positive even when the painful situations aren't so challenging – such as not getting ice-cream, being caught in a traffic jam or stubbing our toe - it takes great effort in reprogramming and training the mind. How much more so for people faced with real trauma, abuse, clinical depression or chemical imbalances. In these cases, the one to blame is the person or condition inflicting the pain, not the victim. The more severe the pain, the harder it is to relate to it in a positive way. It could take years of therapy, healing and/or medication to overcome these challenges. In the end though, the healing will always come when the energy has been released and the mind has come to terms with the pain and can see it in a more positive way, for example by turning the pain into purpose, eg to help others who are going through the same thing. A key book about healing from trauma is Waking the Tiger by Peter Levene.

circumstances or use them to grow and connect. [20] In this way, we are taking full responsibility for writing our own life stories - and not giving the power of our happiness over to things that are out of our control.

In order to attain these levels, some sort of meditation practice is essential. The primary aim of meditation is to tame the mind, to stop it jumping around so fast and uncontrollably, to observe it, focus it, calm it down and taste the stillness hiding beneath. The first thing one notices when starting to meditate is that we are hardly in control of what goes on in our heads.

There are, generally, two types of meditation:

• Passive meditation is the exercise of quieting the incessant chatter of the mind. One way to do this is focussing all awareness on something in the present moment; for example the breath or one of the five senses. In Jewish meditation this is called *hashkata* - quieting. Another way which often goes hand in hand is *habata* - non-judgmentally observing all thoughts and emotions that arise and letting them pass through the consciousness without reaction or resistance. After a while, the breaks in between the thoughts get longer as the focus gets more consistent and effortless, at which point it feels like the deep peace you experience when the baby finally stops screaming and drops off to sleep at three o'clock in the morning. We've all experienced this sublime state naturally for a few moments, perhaps when listening to classical music or the birds singing in the park or feeling the sun on our face on a beautiful spring day. Life is often like treading water, with constant

---

[20] This doesn't mean we won't feel pain. When someone dies we don't just say "this too is for the best" and carry on as normal. We sit and mourn for seven days. Even though we know it's for the best it doesn't mean it's not sad and painful. It's just that the pain doesn't turn into uncontrollable suffering.

waves of challenges, emotions and responsibilities rolling towards us. So much of our energy is taken up with thinking that on the rare occasions when we rise above our thoughts (even for a short while) it is like withdrawing onto the beach, finally resting, finding serenity and recharging. It helps us find the space to be more aware of what is going on in our minds, without drowning in the thoughts. It's amazing to realize that serenity isn't something we need to strive for some time in the future, rather, when the mind is clear and calm, we experience the real peace and tranquility that is always there waiting to be revealed, like the clear blue sky which is always there, no matter how many clouds are covering it.

• Active meditation involves using the mind in a focused and controlled way to get deeper understanding into things, create certain thoughts, feelings and beliefs and approach life in a more positive and empowered way.

An example of this in Jewish teachings is Hitbonenut – contemplation. Contemplating all the gifts we have in life arouses gratitude, contemplating the beauty and majesty of nature arouses awe, contemplating death (in the right, healthy way) arouses purpose and vitality.

It was a great surprise and pleasure for me to find that Judaism has a rich and deep tradition of meditation and spiritual practices for increased awareness and self-mastery. This is not only in kabbalistic literature. The Talmud[21] teaches that the early pious masters would take an hour before praying to help with their intention in prayer, to which the Rambam says that they took that hour in order to 'l'hashchit machshavotam' – to quieten their thoughts, a classic passive meditation before stepping into the active meditation that is Jewish prayer.

---

[21] Brachot 30b

The idea of Kavana (intention) - how we direct our thoughts and attention while praying, doing mitzvot[22] and generally in life is of such huge importance in halachic, Talmudic and Chassidic writings that Jewish law rules that a mitzvah does not really count unless it is done with the right focus and intention. Just going through the motions is not enough.

A man went to the Kotzke Rebbi[23] and told him that he has all sorts of thoughts when he prays. The Kotzke told him that there's nothing he can do for him. The man left a little dejected and returned to his village. On arriving home he saw some men walking in and out of his house taking his furniture out and leaving it on the side walk. The chassid freaked out! 'What are you doing? You can't just walk into someone's house without permission!' The men apologized and told him that they were just following orders – The Kotzke Rebbi had told them to do it! The chassid stormed back to the Rebbi and started shouting ''Not only didn't you help me, you then get people to come and take things out of my house!'' The Rebbi inquired as to why he was so upset about it. 'Because it's my house and no-one can come in my house without permission!' The Rebbi smiled. 'This is your house too' he said pointing at his head, 'no thoughts can come in without your permission.''

So how do we start implementing this practically in our lives? Can we learn basic meditation techniques without leaving home for a ten day silent retreat or sitting crossed legged in a cave somewhere? Can we start to consciously choose what thoughts we allow to come into our

---

[22] As we'll see later, there are 613 mitzvot given in the Torah. Mitzvah (the singular) literally means commandment - but our Sages refer to them as connection points, pieces of spiritual advice or channels of light.

[23] Rabbi Menachem Mendel of Kotzke Poland(1787–1859) A Chassidic Rebbi who was fiercely into personal authenticity and living in line with the truth

heads?[24] See Appendix B for some exercises and practices you can implement in order to integrate these ideas into your daily life.

Judaism teaches that we are not our thoughts, especially not the critical, judgmental, negative ones. Essentially, we are the pure consciousness/soul that lies beyond the mind that can watch, conquer, control and use it. What's more is that developing this level of consciousness is not just "a nice thing to do" if we want to. Just like we are not allowed to put pig in our mouths seeing as it is not spiritually conducive for us,[25] there are certain thoughts that we are not allowed to entertain in our consciousness, seeing as they disconnect us from our peaceful, joyful, pure essence. This means that part of following the Torah's guidelines for living our best lives entails becoming people who can consciously observe the thoughts and emotions which arise and start to interact with them from a place of what Kabbalah refers to as mochin d'gadlus – expanded awareness, higher perspective, choosing what thoughts we let go through our minds, rather than identifying with them, getting caught up in them and reacting to them in speech, feeling and action. In a sense, this is what it really means to be an Observant Jew, able to observe thoughts and emotions from a space of higher consciousness, without being defined or controlled by them.

---

[24] Check out www.litmindfulness.org for my 8 week online Jewish mindfulness course.
[25] Eating pig, like many other Torah prohibitions, is not a moral issue. I know people who don't keep kosher and are wonderful, kind people and I know people who eat strictly kosher but aren't very nice. Someone who eats non-kosher food is not considered a bad person. It's just that it has spiritual properties that we need to avoid.

Enlightened Jewish Master   Thailand

It was a few days after my second ten-day silent retreat that I reached a state of 'no-mind,' of being totally in the present moment, the chatter of the monkey completely silenced, and all this totally unplanned and with absolutely no effort. ... for a few seconds at least. I spent much of my free time in Thailand climbing through the jungle, swimming in reservoirs, scrambling along waterfalls and hanging out on the beach. I was on one of my afternoon adventures into the forests that surrounded my village on my way to the reservoir for a meditation session and afternoon swim. As I emerged through the undergrowth ready to settle in, I was met by a loud noise that sounded like a very fat man snoring extremely loudly. I looked towards the source of the sound feeling a bit annoyed that my plans of quiet solitude had been disturbed and my heart and mind literally stopped for a few moments. About fifty feet away from me, raised up to its full height of around one and a half meters, looking me straight in the eye, was a huge snake, hood fully spread in attack mode, growling and hissing like a large dog, ready to deliver a deadly strike. I didn't know a snake could make such a sound. I was paralyzed with fear and for a few moments of ultimate presence there were no thoughts at all in my consciousness. A few seconds later, my mind kicked back in as thoughts rushed through my head; What if it is a spitting cobra who could accurately spit venom from up to thirty feet? What if they hunt in packs and its friends are coming up behind me? If I had been bitten there, I was pretty much done for, in the middle of nowhere, no phone, no chance. For a while it was just a staring match, helpless little Jew far from home, against pretty angry deadly beast, very much at home. Needless to say, it was me who backed off first, walking slowly away, eyes always on my enemy, intensely alert and pitifully afraid. After regaining some composure, I decided I had to get some identifying marks seeing as if I got out of this alive I'd need to describe the snake to my best friend Ben, who knows all about nature, so he could identify it. I took two careful steps forward, possibly a bad move given the situation, and noted the round markings on either side of its throat. I then turned and ran as fast as I could through the undergrowth, jumped on my scooter and sped away from the reservoir, never to return.

I sat on my porch that evening pondering the fact that it takes many, many hours of meditation practice to come to the state of no-mind, yet when faced with exhilarating near death experiences we get there instantly and with no effort. This is why so many people do bungee jumping and skydiving (and even drugs) in an effort to become totally present, free for a while from the constant barrage of thoughts, or just distract themselves through TV and social media. The difference is that the ultimate presence that is forced upon us by external factors, or the escape we find in entertainment is short lived and doesn't really transform us. The peace and presence that comes from meditation creates lasting insight and real change, bringing us to a higher state of awareness and enabling us to take control of the monkey mind on an ongoing basis. It was about time for evening services, so I made my way to the temple where I took my spot amongst the monks and gently focused my mind on my breath, looking to tap into a place of no-mind, just this time for a more sustained period and without the fear of imminent death. The next day word came back from Ben. As I had suspected, I'd been face to face with one of the world's most deadly snakes - a king cobra.

I sometimes wonder what Tan Medhi[26] would think if he could see me from his place next to the peaceful pond in Suan Mokh, at that moment fifteen minutes before Shabbat candle lighting on Friday afternoon,

---

[26] My monk friend in the temple in Thailand

when the kids are crying, the guests are arriving, I'm still not dressed, the floor needs to be cleaned and then someone spills the red wine all over the white table cloth. He may wonder why I gave up the serenity of temple life for the "balagan" of family life, with rent to pay and mouths to feed. Then again, he may realize that it is one thing controlling your thoughts and emotions while cut off from the challenges of relationships, responsibilities and general everyday life, but quite another to overcome your ego while fully immersed in the world, where the test is far greater and therefore the mastery of the monkey mind has to be on the highest level imaginable.

Chapter Four
Laos
Being the Go-Between

One of my favorite books growing up was 'Kim' by Rudyard Kipling. It's about a young English boy who was born and subsequently orphaned in British-ruled India. I was so enamored with the streetwise kid, running across rooftops, doing deals, learning the ways of the world, of European ancestry yet equally, if not more, comfortable amongst the locals. I prided myself on being somewhat like him; living on the edge between locals and travelers, speaking both languages, knowing both cultures - the connector, not fully part of either world.

I used to walk off into the mountains and find a village where I would stay for a few days, helping with the farming and household chores, taking part in village life, festivals and ceremonies, learning local folk songs, having a true experience of the land. Then I'd hit the traveler hot spots to check out the scene, before retreating back into the hills.

Nowhere more did I take on this role than in Laos, not so much by choice, but rather because I was wrong when I thought my Thai bank card would work there. That left me with $40 in cash to get me through Laos for a month, not impossible if I stayed in a village somewhere, but would certainly not allow me to travel around visiting all the places and *taking part in all the activities I had planned to.* As I withdrew my bank card from the ATM machine on the Laos side of the border having had my request for money denied, I felt the spirit of Kim awakening inside me and I relished the idea of relying on my wits and experience to get me through the next month in a new country, with unforeseeable challenges and adventures awaiting around every corner. By the end of my excursion, I could certainly confirm for myself Rudyard Kipling's wise words in Kim;

"This is a brief life, but in its brevity it offers us some splendid moments, some truly meaningful adventures."

I arrived in Luang Prabang, Laos having taken the two day slow boat ride down the Mekong River from the northern Thai border. The Mekong winds its majestic way from its source in the Tibetan Plateau through China, Burma, Thailand, Laos, Cambodia and Vietnam before emptying out 4,350 km later into the South China Sea. All along her banks, which are framed by mountains and jagged rocks, families go about their daily lives which center almost totally around the river, washing themselves, fishing, playing, doing laundry, collecting water for cooking and using it to move around between villages for bartering purposes. It was not uncommon to see six year olds carrying their two-year old siblings on their backs, filthy and in rags but with huge smiles, sparkling eyes and the joy of childhood freedom shining on their faces. Once again, I became acutely aware of the fact that many poor people in the third world seemed to be more content than many wealthy people in the 'developed' world - with all its privileges.

The language in Laos is very similar to Thai, just a few subtle changes and nuances which I picked up on the boat ride. As had become my tradition, I learnt a couple of local folk songs as soon as I could, seeing as this was the best way to get a deeper insight into their lives and culture, endear myself to the locals and open up a more intimate and authentic relationship with them, helping me be accepted as 'part of the tribe'.

On disembarking, we were met with a truly enchanting scene. The cultural capital of Laos and UNESCO[27] Heritage Site, Luang Prabang is a vibrant city, a fusion of traditional Lao wooden houses, colorful Buddhist temples and French colonial buildings from the 18th and 19th centuries, all surrounded by lush forested mountains home to picturesque waterfalls and ancient caves.

My first stop was a family owned guest house where my banter began – 'I'm not a real farang, I may have the body of a farang but I have the heart of a Laos person! (cue the folk song, much to the delight of the family) I'm not a rich western tourist, I live in a small village in Thailand, just a normal poor man. I'll help you out, what work do you need me to do?' Eventually, the mother had an idea. Many Thai and Laotian boys become novice monks for a few weeks or months to gain merit for their family, as something to do in school holidays or because they have become orphaned or been abandoned by their family. They shave their heads, don the orange robes and enter the temple where they keep the Ten Precepts[28] and learn Buddhist philosophy, meditation and some secular subjects. Often, they don't take it all that seriously, for many of them it's just a tradition they (resentfully) have to go through. The mother thought that maybe having a farang teacher who could

---

[27] The United Nations Educational, Scientific and Cultural Organization
[28] to refrain from destroying living creatures, to refrain from taking that which is not given, to refrain from immorality, to refrain from incorrect speech, to refrain from intoxicating drinks and drugs which lead to carelessness, to refrain from eating at the forbidden time (i.e., after noon), to refrain from dancing, singing, music, entertainment, to refrain from wearing garlands, using perfumes, and beautifying the body with cosmetics, to refrain from lying on a high or luxurious sleeping place, to refrain from accepting gold and silver (money).

speak English and Lao and was familiar with Buddhist concepts and practice may be what the boys needed to inspire them to make the most of their enforced monastic existence. So that's how I became a Rebbi for the novice Buddhist monks in one of the main temples in Luang Prabang! I taught them English, Buddhist philosophy and meditation and, boys being boys, showed them a thing or two with a football at break time. In the afternoons I'd join the other farang hanging out by the river, towards sunset I'd walk up into the hills to relax into a meditative state beyond the hustle and bustle of town life, and then make my way 'home' to the guesthouse to share dinner and stories with my new family. I soon found my routine and once again, rather than being a tourist just visiting a place, I felt like I was really living there, having an authentic experience with the locals and hopefully leaving a positive mark on all those whose paths I crossed.

After a week, I decided it was to time to head for Vang Vieng, the main backpacker hotspot in Laos, boasting captivating untouched scenery, plenty of cafes and chill out places and the famous river tubing tour which involved driving to the drop off point ten kilometers up the Nam Song River and floating all the way back down on a huge tractor tire. On the way, locals on the river banks would throw out ropes and pull you in to partake of local food, have a go on huge swings and ziplines from which you could catapult into the river and consume a fair amount of alcohol and other intoxicating substances. Although you can still go tubing there, the crazy backpacker party was brought to a halt in 2012 as the Lao government was forced by international pressure to crack down on the river side bars, as several travelers, unsurprisingly, didn't make it back to town alive.

It was in Vang Vieng that my inner Kim really came out as I played the go between for the locals and transitory backpackers who often made up over 60% of the otherwise sleepy town's population. The deal was that if I could get some farang to stay in their guest house, go tubing with their organization or eat in their restaurant, I would be rewarded with all these things for free as my pay. So it really ended up being a win-win-win situation. The travelers got discounts on their rooms, activities or food, the local owners got more business, and I got free room, board, tours and adventure activities to boot.

However, still driven by the quest for self-mastery and enlightenment, every morning before sunrise, as the rest of the farang slept off their hangovers, and every evening before the cafe where I worked started filling up again, I'd make my way to the local temple to join the monks in their trance inducing chanting, sinking into the now familiar feeling of inner calm, detached from the physical world with all its challenges, beyond body consciousness, going deeper and deeper into states of one pointedness and no mind, a little Jewish Buddha in the making ...

"Drop by drop the water bucket is filled, likewise the wise man, gathering it little by little fills himself with good."
- Dharmapada

"All Divine service depends on the fixing of character traits"
- Vilna Gaon, Even Shlema[29]

According to Buddhist tradition, Siddhartha Gautama (later referred to as the Buddha) was born in Lumbini, in what is modern day Nepal,

---

[29] The Vilna Gaon, Rav Eliyahu (1720 - 1797) was a genius from a young age and became a major authority on Talmud and Jewish law.

around 2563 years ago. He was a prince who lived in luxury until one day, on a walk outside the palace grounds, he came into contact with the reality of old age, sickness and death. This set him off on his search to discover the meaning of life and to bring an end to the cycle of birth, suffering, death and rebirth. A full look into Buddhist philosophy is far beyond the scope of this book, but in brief the Buddha taught the path of overcoming Dukkha, the un-satisfactoriness of life. According to Buddhist philosophy, everything in existence is anijja – impermanent and shunyata – empty of intrinsic existence and meaning. It is the fact that we get attached to things in this world, especially our own ego, which causes us suffering. He taught that even our ego is an illusion and that really there is anatta - no underlying self, no individual eternal soul and by extension of that not necessarily an infinite Divine Being Who created and controls it all.[30] We suffer in life because we either crave things we don't have, or we are bound to get upset when the things and relationships we do have come to an end. By realizing that really there is no 'Me' and no 'Mine' we break the attachment and craving after things of this world. If there is no 'I' then there can be no feeling of 'I' am unhappy. Of course, the philosophy is much more complex & deep than I have been able to do justice to in this one paragraph, (not to mention the fact that there are different schools of Buddhism which teach slightly different things) but will have to suffice as a general overview for now.

Although its core philosophy is very different to that of Judaism, there are many insightful teachings and practices that the paths share. One of these is the fundamental importance of the development and perfection of character traits. Buddha identified the three core negative traits or defilements (kilesa) as lobha, dosa and moha – greed, anger and

---

[30] Buddha never spoke about the origin of life and creation as he believed we can't get the answer to those questions. There's a debate as to whether he was an atheist or agnostic

ignorance. From these, all other negative traits spring – jealousy, hatred, lust, doubt, arrogance and so on. He also taught that opposite these defilements, there are four main traits which must be cultivated: Metta, Karuna, Mudita and Uppekha – loving kindness, compassion, sympathetic joy and equanimity - or Chesed, Rachamim, Sameach B'Chelko and Histavut in Hebrew. It is through the realization that there is no ego, no 'me' and 'mine' that one can wipe out the negative character traits and develop the positive ones, thereby attaining Nibbana – enlightenment; an egoless, unattached state, leaving one free from the suffering which accompanies life. In fact, the word Nibbana, actually means 'quenching,' meaning to put out the flames of these passions and wrong notions.

Although the emphasis on these character traits is somewhat similar, the Jewish path to attaining this is almost diametrically opposed to the Buddhist one. Rather than saying there is no Self, the Jewish approach to character refinement is that by actually connecting to, identifying with and manifesting our true Self, our pure spiritual essence, we can manifest our perfect, pure nature in the world. Every morning we say 'Elokai, neshamah she'natatah bi tehorah hee' – 'My Gd, the soul you have given me is Pure' and the prophet Yeshayahu[31] taught that "Amech kulam tzaddikim" – We are all fully righteous, enlightened beings. It doesn't say that we *will be* Tzaddikim, rather that we are already Tzaddikim – right here and now. The only issue is that we don't often think, feel, speak or act like one. Yet, just like an elephant who dresses up like a chicken, makes chicken noises, acts like a chicken and eats chicken food is still really an elephant; when we strip away the thoughts, speech and actions that are not in line with our soul, we reveal our essential enlightened, connected, healthy, pure Selves. This means that rather than creating health, connection, holiness and joy, our job is to strip away everything that is covering it up, revealing that it is already

---

[31] Yeshayahu 60:21

there. The first words in the Torah are 'Bereishit bara Elokim et hashamayim v'et ha'aretz' - in the beginning of Hashem's creating the heavens and earth. The word Bara, literally translated as 'create,' has the same root as the word bari, meaning healthy. The Torah is telling us that the natural state of the universe and everything in it, including us, is totally healthy and in tune. It's just when we start to misuse and abuse it that we move away from this state of being. It turns out that Judaism is not a path to reaching enlightenment, it's a path to revealing that you are already enlightened.

On a simple level the Rambam teaches that the mitzvahs were given to; *'subjugate one's evil inclination and improve one's character. Similarly, most of the Torah's laws are nothing other than advice from the Creator to improve one's character and make one's conduct upright.'*

There are many examples of this in the Torah. In the portion of Mishpatim, we are told that when we see an enemy struggling we must surely go and help them, not letting our prejudices (even if they are well-founded) get in the way of doing the right thing. We must abandon our hatred, resentment and grudges, our envy, anger and hurt, in order to rise to the level of manifesting G-dliness in the world.
I used to find it strange that the Torah commands us to feel a certain way. Telling us to put on tefillin, light candles, eat matzah, I get. I can do that even if I don't feel like it. But to command us to feel something - to love Hashem,[32] to feel more joy at certain times, to not bear a grudge

---

[32] Hashem literally means 'The Name.' It's how we refer to G-d. The word Gd has lots of connotations and images to it - an angry man in the sky who wants to punish us. Hashem is a more gentle way to refer to the Infinite, Intelligent, Loving Consciousness from which the world was created.

- just doesn't seem to make sense at all. How can one be commanded to feel something? What the Torah is teaching us is that we can indeed be in control of our emotions; not suppress or deny them, but to fully experience them and reshape and develop them in a healthy way. We have negative thoughts, speech and actions, but that doesn't define us. We are here to change and refine these aspects of ourselves, not resign ourselves to them.

Further to this, the Kabbalists teach that the soul is actually a 'piece' or aspect of Hashem, a finite manifestation of infinite consciousness. Therefore by identifying as, connecting to and experiencing our soul, we partake of the Divine perfection. In the Torah, there are 613 mitzvot. Literally meaning 'commandments,' our Sages have variously referred to them as 'connection points,' 'pieces of spiritual advice,' and 'channels of light.[33]' The mitzvahs are there to feed and connect us to our soul (if done with consciousness and joy rather than out of fear or habit), thereby lighting up our path towards our goal in life.

There is an amazing story in the Torah which teaches that Hashem appeared to Avraham while he was sitting at the entrance to his tent[34]. Unlike every other place in the Torah in which Hashem appears to someone, the verses don't tell us what Hashem said. It appears that Avraham and Hashem weren't even communicating, they were just sitting in the Highest state of Divine meditative communion (similar to what the Buddhists call Samadhi). Then, suddenly, Avraham spots some dusty travelers, tells Hashem to hang on for a second, and rushes to wash their feet and feed them.

Seems a bit rude to me. If the point of doing the mitzvah of greeting guests is to be on a high spiritual level and connect to Hashem, then Avraham was already at the highest level and didn't need to do a mitzvah to connect.

---

[33] Orot - the Hebrew word for Lights has the numerical value 613.
[34] Bereishit 18:1

Yet we see that's exactly what he did. The Talmud[35] uses this story to teach that it is greater to do the mitzvah of welcoming guests than to welcome even the Divine presence (shechina) Itself.' Rav Noach Weinberg, taught that from this story we see that the Jewish ideal for us in this world is actually not to be *with* Hashem in some sort of spiritual bliss, rather, it's to be *like* Hashem. Just like Hashem is a perfect being containing all the love and goodness possible,[36] so too we must strive to manifest that in our lives. The more we are *like* Hashem in this world, the more we'll be *with* Him when the soul leaves the body and re-enters the spiritual reality on the other side of this physical life.

So how are we going to start implementing real growth in the area of character traits? Rav Yisrael Salanter[37] himself said that to change one trait is harder than learning the whole Talmud. We may not be able to completely uproot them, but we can certainly begin to stop them playing such a strong role in our lives. A girl once told me she hates herself. When I asked why, she listed a few negative character traits. I said "I hear that, but those are your negative character traits you need to work on, they aren't you. If your kids have negative character traits you won't hate them, so don't hate yourself for yours!" Never beat yourself up about negative character traits. That is a trick of the yetzer hara (evil inclination) to disempower you. No-one ever improved by beating themselves up. King David said Ivdu Et Hashem B'simcha – spiritual work has to be done with joy – putting ourselves down never works.

See Appendix C for some exercises and practices you can implement in order to integrate these ideas into your daily life.

---

[35] Shabbat 127a
[36] I know it doesn't feel like that sometimes; we'll speak about it in a later chapter
[37] Founder of the Mussar Movement focussing on personal development and perfection of character traits

How does this fit into our picture of developing real Simcha (joy) in life? The only thing standing in the way of us feeling our innate joy is our own negative character traits. You may say 'No, it's my boss.' Truth is that if you had more patience or self confidence you'd deal with your boss. It could be your upbringing - yet there are many people who have become great by working through and healing from their challenging start. The more we take responsibility to fix our own character traits, choosing not to put blame and be a victim[38] the happier we become. We see once again that searching for happiness outside of us is futile. All it really takes is a shift of attitude to life, the desire to change and the tools and discipline to actually implement the ideas.

It was actually in this area of character traits that I started to question the coherence of what I was learning through the Buddhist sources. I was wandering around the temple grounds one day in Vang Vieng, Laos, contemplating the teachings: on the one hand the importance of nurturing love, compassion and sympathetic joy; yet on the other I was meant to not get attached to anything of this world.

The question arose - How can I be truly loving and compassionate if I am also meant to be detached? Surely compassion involves some amount of attachment to the one I feel compassion towards and love involves some sort of intrinsic connection. I realized that, far from being detached, it's through knowing that we are deeply and intrinsically connected that these feelings can arise. Although I asked my masters and searched through the many answers to the paradoxical idea of detached compassion, none of them really sat well with me and I realized that I still had a lot of searching to do if I was to find a fully coherent path to cultivating the character traits I needed to develop along the path towards true emotional and mental Enlightenment.

---

[38] even if this would be justified

After a while in Vang Vieng I decided it was time for a real adventure, off the beaten track and away from the backpacker trail. I took a bus up north to its last stop, got down in a small town and asked around for the way to villages in the mountains. I packed a few provisions and set off, not sure where the trail would take me or what lay in store over the mountain passes. I used to say that you are not really living unless you have dirty feet. As I looked down at the blackened toes sticking out of my sandals, I smiled contentedly. Nothing felt as good as being on the solitary trail to who knows where. I walked through a couple of villages where all the kids ran out to say hi and ogle the strange apparition walking through their neighborhood. None of the villages had running water or electricity, the only water source was a brown river that they bathed in, drank from, went to the bathroom in and used for cooking. As I left each village the kids would escort me on my way until their parents called them back to finish their chores or have something to eat. After walking about twenty kilometers up into the hills I finally arrived in a village about two hours before dark and decided it would be best to find a place to settle in for the night. Curious faces popped out of the hut windows as word got out that a farang was in the village. I found out later I was the first foreigner most of them had seen. At 5 foot 8, height isn't my greatest asset, yet when I arrived in the village I stood a good head above the tallest person there. After partaking of their diet of sticky rice and stringy mushrooms from the river for over a week, I realized that malnutrition had really stunted their growth.

The chief invited me into his hut and soon word had gotten out that there would be a gathering at his house for all the villagers. The whole community crowded round as I sat on the porch and addressed them. I'm not sure they really understood what I was showing them when I pulled out my well-worn map of their country on which their village obviously didn't appear and seeing as my camera was not digital they had to remain a little confused as to what I had done to them with the little mental box I carried with me. We sat drinking the rancid local rice wine through bamboo straws and the villagers began to ask me questions. 'Do you have mountains in your country?' 'YES, big mountains' said I, melodramatically, throwing my arms in the air. 'Are there many farang in the world?' 'YES, many.' It carried on like this for about an hour before it started getting late and the crowd began to disperse. That night as I settled into my sleeping bag on the wooden floor of the chief's hut, bag under my head as a security measure against thieves as much as for use as a pillow, I started pondering life. These people basically never left the village. They had no education, health care or passports. They spent the majority of their waking hours farming and collecting water and wood in order to cook the food they farmed, so that they could eat, to stay alive, to farm and collect wood and water to cook to eat ... it seemed like an almost meaningless cycle. In their free time they would sit around drinking and telling stories, playing games, and other general entertainment to fill in the time. I wondered if they ever stopped to think about the purpose of it all. Then it hit me that it isn't that much different where I came from! Many people are in a cycle of going to work that doesn't particularly fulfil them and then spending their free time entertaining themselves in some way, without ever really stopping to ask themselves what the point of it all is. The quality of life is definitely better and more comfortable, the entertainment is more entertaining – but in the end it's a similar cycle of just working to be alive - and get occasional temporary pleasure. It became clear that the answer to my search for purpose was not to be found in the world I had come from, nor the almost diametrically opposed lives of the people in the village. I still had a long way to go on my warrior's journey to discovering and fully living in line with the true purpose of life

Chapter Five
Korea
Black Belt on a Paradise Island

*S*now balancing precariously on the outstretched leaves of palm trees lining the coastal roads, ripe deep-orange citrus fruit with glistening ice-caps thriving in mikan[39] groves, Halla-san, the once fiercely active volcano which still dominates the island, swathed in a frosty white carpet, frozen streams running through forested valleys like paths left by snails across the lawn, Christmas lights and overly decorated trees adorning the shop fronts of Seogwipo-City, school children wrapped up in hats, scarfs, gloves, leg warmers and earmuffs, carefully hurrying home on the icy pavements – Jeju-do, which was my home for just over a year, was magnificently showing that even in mid-winter it is one of the most spectacular places in the world.

After nearly a year and a half in Thailand and Laos, I felt the strong drive to move on to fresh pastures, face new challenges and adventures, experience an unfamiliar culture with different scenery and ultimately to start seriously pursuing my dream of training in martial arts. One day, while browsing a TEFL[40] website, I was attracted by a headline - 'Come teach English on a Paradise Island.' I contacted the Hagwon (after school academy) that had posted the vacancy, sent in my CV, had a phone interview, was offered the job and within two months arrived at my new home on 'Honeymoon Island.'

---

[39] A type of sweet mandarin found mainly in Korea, Japan, China and South Africa.
[40] Teaching English as a Foreign Language

Lying 100 km off the coast of South Korea at the convergence of the Yellow and East China Seas, Jeju-do really is a paradise island and possibly the most phenomenal place I've ever lived. Dominated by Hallasan, the tallest and one of the three most sacred mountains in Korea, the island, which was created entirely through volcanic eruptions, is home to a stunning array of natural wonders. It boasts one of only eleven waterfalls in Asia that falls directly into the sea, over 360 oreums (mini-volcanoes), one of the longest and best preserved underground lava tunnels in the world and a mysterious road on which bottles and cars out of gear (seem to) roll uphill. Then there are the ideal beaches lining the coasts, intriguing rock formations called Dol Hareubang – three-meter tall carved statues of unclear origin, said to give the power of fertility and protection from demons; parks and gardens full of thousands of species of flowers and plants, waterfalls hidden in the depth of the forest which covers over ten percent of the island and one of the biggest Buddhist temples in the world. Jeju-do, although officially part of Korea, has its own distinct culture, dialect and very rich body of local myths and legends.

After taking a couple of days to settle in, I walked into Jong Il Hagwon for my first day of work and was met with a rather disturbing scene. A chubby eleven year boy was in the push-up position with Mrs. Park, the principal, standing over him shouting, half a pool cue raised in her hand. She turned around when she heard the door open and her eyes lit up when she saw who it was. She gave me the sweetest smile you could imagine through her bright red lipstick-smeared mouth, turned back around and gave the boy one firm strike on his behind with the pool cue, probably much lighter than he would have received had I not walked in just then.

She then invited me to join her in the office, proceeded to explain the rules of the school, gave me a timetable, textbooks, board markers and my own half a pool cue which she assured I could use whenever I saw fit. Unfortunately, I very often felt the need to use it, however I took my frustration with the kids out on my desk instead, which made a loud enough sound to get everyone's attention without leaving red marks on their bodies.

In my free time, I'd cruise round the island on my motorbike with some of the other English teachers, enjoying the beaches and waterfalls, hiking the forest trails, playing tennis, visiting various temples and shrines, learning martial arts and climbing Hallasan, the volcanic centerpiece of the island. There were teachers from Canada, Australia and America and we formed a tight knit group of expats. Many of them were Christian missionaries and I became very friendly with a pair of Mormons, once attended a 'cell meeting' where some passages from the new testament were discussed, hands were held and songs were sang, went to a few great concerts at a hip progressive church in the neighboring town where some of my colleagues from work were performing and I became top scorer for the local Presbyterian Church football team. Seogwipo had been one of the host cities for the 2002 football world cup, with the likes of Germany and Brazil gracing the turf at its new $120 million stadium. Football had become a craze on the island since then and new teams were springing up all the time, not least of all in the church league. I got connected to the team through one of my students and once again created strong friendships through the 'beautiful game.' My career came to an abrupt end when the captain of the team excitedly announced that there was an island wide church tournament and the final would be played at the world cup stadium! He came up to me and told me that to qualify to play in the team I'd have to come to church every Sunday. That same inexplicable drive that caused many non-religious Jews to give up their lives rather than bow to an idol or eat pig kicked in and I told him that Jews can't go to church (even though I used to love sitting in the meditative atmosphere of churches) and so I'd have to sit out of the tournament, hence my own little Sandy Koufax[41] moment.

Although Jeju still has a strong local culture and traditional feel to it, Korea, unlike Thailand and Sri Lanka, is definitely a 'First World' country – defined by Wikipedia as 'democratic, capitalist, economically stable, technologically advanced and with a high standard of living.' Going in the face of those who say that money is the root of all evil, I could see lots of good that came from it - health care, education, welfare for more disadvantaged people in society, much lower rates of poverty, unemployment and homelessness, to name but a few. Yet on the other hand, contrasting the 'developing' countries I had just come from to the first world one I was now in definitely highlighted the corrupting effect of consumerism on individuals and society as a whole. There was markedly less emphasis on community and familial closeness and even though the standard of living was higher, there was far less contentment and happiness amongst the people. Along with this drive for modernity came another consequence that although the older generation were still deeply imbued with a sense of tradition and love of the local culture, the young had been completely taken over by American culture. Hip hop and baseball have taken over from minyo and taekwondo as the most popular music and sports, and Christianity was the new religion for the spiritually progressive youth.

Of course that didn't stop me. As with everywhere else I went, I was fascinated with local culture, history and tradition. Once I had perfected Jeju-satori (the local dialect) I refused to speak mainland Korean when there was an equivalent local word, I still spent many hours meditating in local temples and as a punishment for those who misbehaved in class, rather than wielding my half pool cue, I made them stay behind after class for myongsang jikan – meditation time. At first my students found

---

[41] Legendary Major League Baseball Hall of Famer in the 1950's and 60's who opted to sit out of Game 1 of the 1965 world series seeing as it fell on Yom Kippur.

it amusing (or annoying), yet after a while I could see they were starting to be infused with a love of and pride in their island culture and when they saw me on the street they'd say 'Yogie Jeju-Saram Onda' - Here comes the Jeju Man!

For me, coming back to the 'developed' world after spending a couple of years in Sri Lanka and Thailand, really highlighted the fact that 'developed' is in many ways a terrible misnomer. We may have progressed economically and politically, but society seems to have gone in the opposite direction. It was clear that the emphasis on capitalism and hedonistic pleasure is not creating better, happier or healthier individuals. I started to wonder whether there was any way to have the best of both worlds - finding a balance between having financial stability and high physical quality of life, while still living with healthy and wholesome values which promote the physical, emotional, intellectual and spiritual wellbeing of individuals and society as a whole.

It was only several years later, while learning Rabbi Dessler's classic work, *Strive for Truth*, that I found the answer. All it would take is a small shift in human consciousness and attitude, one which would create the balanced and healthy individual and society we're all hoping for. Until then I'd just have to continue playing the rather absurd role of middle class Westerner with Thai villager values living as traditionally as I could on a Korean Island where the locals were striving to become Westernized! It was a role that I relished at the time, yet even then I knew it couldn't be my final destination.

"Whoever dies with most toys wins" — Malcolm Forbes.

"You shall repeatedly give (to the poor), and your heart shall not be grieved when you give to him" - Deuteronomy 15:10

Notwithstanding all the growth, beauty and human goodness we find in the world, something has gone horribly wrong. Society is functioning on an extremely low level of consciousness. There is war, an absurdly unbalanced distribution of wealth and resources, abuse, physical and mental illness and we are destroying the planet by polluting the seas and chopping down the forests. What's going on here? Where did it all go wrong?

The problem really begins with the fact that people are chasing happiness in the wrong place. We have an innate drive towards pleasure and away from pain and have developed the erroneous belief that the thing that will bring most fulfilment and pleasure in life is material wealth. This is based on the fact that to some extent material possessions certainly do increase happiness levels. If you are homeless and cold, acquiring a small apartment and getting some hot food will certainly make you happier. If you have to ride your bike up lots of hills to get home from work after a hard day, buying an electric bike will give you great pleasure. Yet the truth is that once we have the basics, adding material possessions will never actually fulfil our emotional, intellectual and spiritual needs. In fact, quite the opposite is true. Research has shown that this focus is what causes most mental and emotional problems, seeing as wanting more than we need, and comparing ourselves to others means we will never feel fulfilled, satisfied and significant. A recent report shows that up until an income of $75,000 a year happiness levels increase, but from then on, having more money makes absolutely no difference.[42]

Unfortunately acquisition and material gain have become the highest values in society. Take a second now to imagine what a very successful person looks like.

If you are like most people, you pictured someone with a nice car, house, suit, haircut. Success in western secular society, whether anyone personally believes in this or not, is usually defined by superficial, external values; what you have, how much you have of it and what you, your things and your partner look like. Magazines, TV shows and movies glorify wealth and physical beauty and the education system is geared towards building a career rather than a good character[43]

Another major issue with this is that most people have very unhealthy self esteem. If my sense of self is based on external recognition based on superficial values, I'll either have low self esteem if I don't have these things, or arrogance (which is another form of low self esteem because I'm still needing external validation which never gives peace of mind) if I do. Too many people nowadays are down on themselves with their self-critical voice running wild. Very few people have a healthy sense of self worth. The truth is that you must give yourself permission to love yourself for no reason, even though you aren't perfect. I love my wife and kids for no reason other than they are my wife and kids. They don't need to perform or do anything to receive my love, and there is nothing they can do to make me not love them. The same goes for ourselves. We don't need a reason to love ourselves and nothing we do should make us stop loving ourselves. Everyone is worthy of love. Judaism teaches that you are intrinsically worthy and loveable. Healthy self-esteem is based on what we like and respect about ourselves; our good choices, values, achievements and character traits and how we are

---

[42] Study from Princeton University reported in Time Magazine Monday Sept 6th, 2010
[43] See Ken Robinson's excellent TED talk about how schools kill creativity.

using them to help others (nothing to do with your car, legs, job, social media posts).

In his 2010 documentary I AM, Tom Shadyac[44] points out that in the Native American culture someone who hoarded things was considered mentally ill. He goes on to demonstrate that hardly anything in nature takes more than it needs. The redwood tree drinks the amount of water it needs to stay healthy, the lion kills only one deer at a time. Humans are one of only two things on the planet that is crazed about hoarding and acquiring beyond our needs. The other one is cancer. Cancer destroys everything around eventually killing off the organism that is keeping it alive. It appears that we are doing the same to our planet. The Torah[45] says "Hashem put Adam in the garden to work and guard it" and the midrash[46] comments "Be careful not to spoil or destroy My world for if you do, there will be nobody after you to repair it," and yet we are going a long way to destroying it by abusing it and taking from it in an unsustainable way.

It's important to note that according to Jewish teachings there is nothing wrong at all with being rich and having nice things. If you study and work hard you may deserve possessions to enjoy and it is noble to provide for your family. Some philosophies teach that we shouldn't have money - it corrupts you, makes you greedy, causes worry, makes you arrogant, makes you partake of things you shouldn't partake of. Although, unfortunately, this is often the case, it doesn't have to be like that. Many of our greatest sages (including Moshe Rabbeinu and Rebbe Yehuda HaNasi) were very rich. Many mitzvahs such as tzedakah

---

[44] Director of Ace Ventura, Bruce Almighty and other films, who after having a near death experience had an epiphany about life and made a documentary about what he discovered.
[45] Bereishis 2:15
[46] Koheles Rabbah 7.28

maiser,[47] hachnachat orchim[48] and pideon shvuyim[49] are accomplished with money. So money itself clearly isn't the problem. The real problem comes when people start making money their top value at the expense of other people and the world around them, defining themselves and their self-worth by it. The World Watch Institute reports that the amount of money spent annually on cosmetics in America is $8 billion and that an extra $9 billion a year would provide clean water and sanitation to all people in all developing nations throughout the world. Consumerism is consuming us and the world around us.

Ultimately, this ego based, self-gain hierarchy of values is not only causing personal suffering and environmental catastrophes; it also causes abuse and suffering between people. When we value our own wealth, power and physical pleasure over that of other people, we will inevitably cause suffering in the world.

So what is the answer? How can we start healing and stop harming the planet and the people on it? Surprise surprise, one of the key paradigms for perfecting ourselves and the world around us, as shared most famously by Rav Eliyahu Dessler, is the absolute reverse of the consumerist acquisition trend we see nowadays. Rav Dessler divides the world up into Givers and Takers and urges us all to move strongly away from being a taker, an acquirer, a consumer, feeding our own ego desires, towards becoming a conscientious and passionate giver. He even identifies the desire to take as 'the root of all evil in the world,' and from what we learnt above we can see how this plays out. We need to start moving from 'What's in it for me?' to 'What can I do for others?'

One of the "less good" people in the Torah was a guy called Korach, who started a rebellion against Moshe. The Torah portion starts with the

---

[47] Giving 10% of your income to charity
[48] Having guests
[49] Redeeming captives

words ויקח קרח –' And Korach took' but then completely fails to mention what he took. Many commentators have given many answers to what he took, but in a sense it doesn't matter what he took. The fact was he was a taker, feeding his ego rather than supporting others was the real problem. His fate was certainly fitting as he was ultimately swallowed up by the earth – the consumer became the consumed.

Success in Torah terms is not based on what you have, it's based on who you are, how much you have perfected yourself in thought, speech and action and how much you are giving to those around you. The Torah is full of admonitions to perfect oneself and help others; ultimately to become a tzaddik– fully righteous person. [50] The Talmud[51] teaches that before a child is born he is administered an oath to "be a tzaddik (righteous) and not to be a rasha (evil)." This means that our area of free-will is in choosing good over evil, perfecting ourselves over feeding our ego and physical desires, being a giver not a taker. In fact, a large amount of Torah law is to do with how to treat others, fairness in business, correcting our character traits and making sure everyone in society is getting the opportunity to also work towards the goals of reaching their potential and helping others.

צדק צדק תרדוף – You shall surely chase justice כמוך לרעך ואהבת – You should love your fellow as yourself, לנקום לא – don't bear a grudge, לנטור לא – don't take revenge, תחמוד לא – don't covet, feed the vulnerable, fight jealousy, hatred and anger, give tzedakah and maiser and so on. When asked by someone wanting to convert to teach him the Torah on one foot, Hillel famously replied 'That which is hateful to you, don't do to others - that is the whole Torah, the rest is just commentary.'[52]

---

[50] There are varying definitions of what it means to be a tzadik. For now, I'm just using the general meaning of a fully righteous person, sage, saint, enlightened one.
[51] Nidah 30b
[52] Shabbat 31a

The Talmud[53] teaches that The Torah itself begins and ends with acts of kindness. The list goes on and on. We see this beautifully manifested in the Jewish attitude towards rights and obligations. Rabbi Akiva Tatz points out that *"Modern society is largely concerned with rights. The highest code of such societies is their Bill of Rights. In striking contrast is the Torah, the Jewish Constitution. The Torah never mentions rights, only obligations!"* [54] Why is this? He goes on to share that rights and obligations always come together; my right to life is your obligation not to kill, your right to property is my obligation not to steal. However, there's a huge difference in which one you focus on. Rights are egocentric, what is owed to me, whereas obligations shift the emphasis to what I can do for others, on being a giver, not what's in it for me, but what I can do for others. Relationships and societies built on this attitude are happy and healthy.

Judaism teaches that there is actually a whole higher reason for and realm of giving. As we spoke about in the last chapter, our lives are about tapping into and manifesting our Gdly nature. So just as Hashem is a giver, it is only by being a giver that we can ever achieve this goal and continue the flow of chesed that comes from Him/Her. Just like water in a pond becomes stagnant if it isn't taking in and giving out, so too we as human beings start to deteriorate if we aren't reaching out to others around us. In Israel there are two big bodies of water - Lake Kinneret and the Dead Sea. They both come from the Jordan River. Lake Kinneret is full of plant and animal life, the Dead Sea, as its name suggests, has no natural life. The difference? The Kinneret takes in water (in the north) and gives out (in the south) - keeping the flow of life going. The Dead Sea on the other hand, takes in but doesn't give out.

---

[53] Sotah 14a – In the beginning Hashem clothed Adam and Chava and at the end He buried Moshe.
[54] WorldMask: The World of Obligation p100 – 106

Finally, our Kabbalistic masters teach that there is a whole spiritual level of giving. Many people will tell you that there is nothing we can give to Hashem, He[55] is perfect so there is nothing He needs. Although this is certainly true on the highest level, our sages teach us that He set up creation in a way that He does, so to speak, in fact need us. In Tehillim[56] that King David tells us 'Give strength to Hashem.' How exactly do we do that? What strength could we possibly give Him? There are two answers to this question. The Midrash[57] teaches that Turnus Rufus, the Roman General, asked Rabbi Akiva 'What is better – the work of Gd or the work of man?' He was making a protest against circumcision – How could you desecrate Gd's handy work? Rabbi Akiva showed him some wheat and some bread and said 'Aren't these (bread rolls) better than the sheaves?' Hashem provides us with the raw materials and asks us to partner with Him in uplifting the world to perfection. We give to Hashem by partnering with Him in bringing creation to perfection. The second level is an even more revolutionary idea. Although Hashem's existence is not at all dependent on our believing in Him, He, so to speak, He has created existence such that, unless He openly reveals HimSelf - which rarely happens, the only way He can be manifest in the world is through our recognition! The source for this idea is in the Torah portion of Chayei Sara. Avraham calls Hashem 'Gd of the heavens and Gd of the Earth'[58] and just a few verses later just 'Gd of the heavens.'[59] How do we explain this? Rashi teaches

---

[55] Why do we generally refer to G-d as He? Gd clearly isn't a man, and seeing as Jewish tradition teaches that women are actually on a higher spiritual level than men it makes even less sense. The mystical answer is that in the act of procreation, the male gives the seed and the female takes the seed and develops into a child. So too in the relationship between Hashem and humans, Hashem gives the seed (wheat/our character traits good and bad) i.e the raw material and we have to develop it into something greater (bread/ self actualisation).
[56] Tehillim 68:35
[57] Tanchuma Parshat Tazria 5
[58] Bereishis 24:3
[59] Bereishit 24:7

that Avraham said that *'Now He is Gd of heaven and earth seeing as I have familiarized Him in the mouth of the people (they speak about Him), but when He took me from my father's house He was Gd of the heavens but not Gd of the earth, for those who lived in the world did not recognize Him.'* Only once Avraham had spread G-d-consciousness in the world, could Hashem truly be said to be here.

It is told that the Chassidic master, Rabbi Menachem Mendel of Kotzk, once asked his students: Where is Hashem?" They seemed a bit confused at the simplicity of the question and answered that Hashem is everywhere. The Rebbe looked at them and said, "No. He is wherever you let Him in."

That we are here as partners with Gd and have something we can give Him (so to speak) is a radical and awesome idea, truly exceptional amongst world religions. Far from being somewhat insignificant sinners trying to ingratiate ourselves in the eyes of Gd, or trying to reach a state of unity with Gd through secluded meditation, Judaism teaches that Jews, once we become givers, are necessary Divine partners in the journey towards the perfection of the universe.

Everyone I know wants to be successful. From the last two chapters we see a paradigm shift in the definition of success. Success, in Torah terms, is not based on what you have, it's based on who you are, how much you have perfected yourself in thought, speech and action and how much you are giving to those around you. When imagining a successful person what should come to mind is not someone with lots of money, rather it is a tzaddik - someone with refined character traits, a good attitude and disposition, someone who is there for others, someone who always does the right thing.

We actually all know this intuitively, as at every funeral the eulogies focus on their good character traits and how much the person contributed to the world around them, not on how many cars they had and how much money they made. What we are aiming for, what we get self-respect based on, what creates a meaningful life and therefore dictates the amount of simcha we have is not how much we make but how much we give. It's not about how we've prospered rather how much we've grown. It's not about looking good, it's about being good. Healthy self-esteem comes from respecting ourselves for how much we've grown and contributed, not what others think of us based on superficial values. Reaching success in these terms is the key to living a meaningful, constructive and happy life.

See Appendix D for some exercises and practices you can implement in order to integrate these ideas into your daily life.

An immensely important thing to understand in this area is that we have two powerful drives. One is to look good and feel good. This comes from our animal side (body/ego) looking for physical and ego pleasure. The other is to be good and do good. This comes from our Gdly side (soul). We want to get out of bed early to go for a run and we also want to press the snooze button, curl over and go back to sleep. We want the chocolate and we want to have self control. We want to shout at someone we love but we don't want to hurt them. Whichever one we feed will become stronger. The truth is that if we chase feeling good and we attain what we desired, we actually end up not feeling so good. Once we stop chasing looking good and feeling good and rather focus on being good and doing good, we'll actually end up feeling really good about ourselves (and looking good in the eyes of others). In this way we live a meaningful and therefore joyful life. This is real success. The beautiful irony is, that rather than by increasing what we have, it's by perfecting ourselves and giving to others that we actually fulfil that innate drive for pleasure in life in the highest way, moving from temporary ego pleasures to higher emotional, intellectual and spiritual ones. Once we achieve this attitude shift, we'll be able to live a balanced life of financial stability and high standard of living, without compromising the values and attitudes which create healthy, happy individuals and societies, just as the Torah has been guiding us towards all along.

It was on Jeju Island that I finally started my real martial arts training. One of my students introduced me to her taekwondo master and from that day on a beautiful master – student relationship developed as I went for private lessons with Oa-Sabomnim (Master Oa) for an hour and a half every morning. He was a very sweet and simple man, born and brought up on the island, fiercely proud of his culture and endearingly interested in mine. After training we'd often go out into nature, jumping into freezing waterfalls, scrambling up mountain paths and would sometimes meet up at night in one of the many fried chicken and beer joints downtown to practice each other's respective language, which was pretty challenging - seeing as when I arrived he couldn't speak one word of English (and when I left he could only speak about three), and my Korean only became really conversational after about three months. It's incredible how two people who love and respect each other and want to connect can connect so deeply even without being able to put together one full sentence.

After a few months, due to lack of students, O-Sabomnim had to close his dojo, so we joined forces with his friend, Kwanjangnim (Grand master). Kwanjangnim was a huge man, well over six-foot tall, former high school national taekwondo champion, and although he had put on a bit of a belly in his later years, he still had the most deadly pair of legs

I've ever had the misfortune to be repeatedly kicked by. Many a time a kick onto the pads he made me hold for him would actually lift me slightly off the ground and send me rolling backwards a couple of feet. Training included jumping through hoops to side kick wooden boards in half, jumping off walls, spinning in the air to kick a board and land all on the same foot, punching out candles without touching them, practicing pumsae (forms) over and over again and georoge (sparring). Looking back, I have no idea how I ever managed to do it all without breaking an ankle or worse. With the daily intensive training, plus an early morning run and meditation, and a gym session after work it took me a year and two months to get a black belt, significantly less than it takes in England, probably because they train far less and the masters would definitely be sued if they treated their pupils like we were treated in Korea.

On the day I got my black belt, Oa-Sabonim and Kwangjangnim took me out for a special meal to celebrate. It was a very pricy and fancy restaurant that served local traditional specialties, playing traditional music, traditional décor and waiters in traditional dress. As we tucked into our meal, downing soju and singing local folk songs at full volume, I reveled in the fact that I'd finally found the perfect fusion of modern opulence and traditional culture.

Yet the next day, while floating in a rock pool of crystal clear water looking up at the cloudless blue sky near my apartment, relaxing on a day off from training, I suddenly felt that there was something missing in my life. I realized that I had everything I could want on my paradise island, apart from one thing - the opportunity to really be there for other people, to give of myself and my time to improve the lives of those around me. Everyone on the island was well fed and educated (or at least had the opportunity to be) and society was running very well on the whole. At that point I decided that it was time to move on, back to the 'developing world' back to the place which on the one hand was so alive and rich and deep in culture, yet on the other struggling so heavily with poverty and other forms of deprivation. I didn't have to think long where the next stop might be, as the time had clearly come to fulfil my dream of visiting the former stronghold of the British Empire, home of the yogis and ascetics striving for Moksha (liberation/enlightenment) in the foothills of the Himalayas or the plains of Tamil Nadu: It was time for me to hit the vast subcontinent of India. When my contract at the hagwon came to an end, I packed my bags, bought a one-way plane ticket and within a week I stepped out of the airport into the stifling heat of New Delhi.

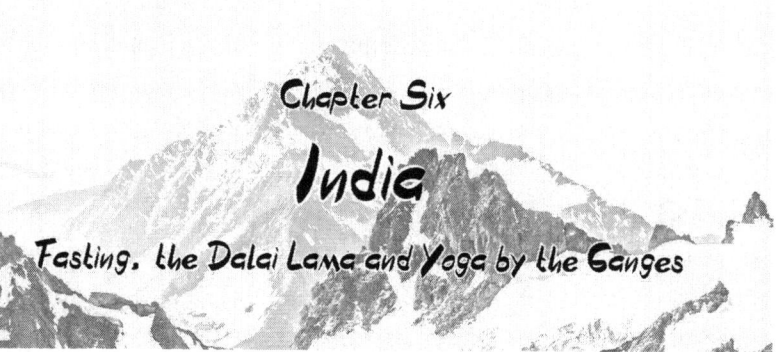

Chapter Six
India
Fasting, the Dalai Lama and Yoga by the Ganges

*A*nyone who has experienced the multi-sensual, visceral, geographical, cultural vastness which is the Indian subcontinent, knows that no-one can really do it justice in one chapter of a book, even the longest one - but I'm going to do my best. By the time I arrived there three years into my journey through the East, I was deeply entrenched in the spiritual search, on a quest for the truth of reality, ultimate consciousness, self-knowledge, the mystical experience, enlightenment – known by the locals as Moksha - Liberation.

Looking back, I'm not sure why I thought my Self may be hiding somewhere in an ashram in India, but like many western Jews, I knew it couldn't be back in the life and values I'd left behind, nor in the religion of my birth which seemed so outdated, rigid and unspiritual.On arrival I went straight to McLeod Gang, a small village half an hour up a very windy road from Dharamasala in the northern province of Himachal Pradesh. It's the main home of the Dalai Lama and the Tibetan people in exile, (and more recently the Israeli backpacker in exile). When I say straight there, I mean after a pretty crazy fifteen hour winding, potholed overnight bus ride from Delhi, which to be fair, by Indian standards, is not actually that long. In the foothills of the Himalayas the air is crisp and fresh and the sky seems to stretch up higher than in other places. Birds of prey circle overhead and the valley stretches miles out to the plains below. Colorful prayer flags flap in wind, Tibetan school children stand on street corners in their immaculate blue uniforms, monks with shaven heads and red robes bustle through the streets talking on their cell phones. Meditation is much easier there, the atmosphere just brings it out of you. Spiritual consciousness pervades the whole place. I settled into a guest house overlooking the valley and noticed that next door was a center where people were paying to do special fasts for seven days. The idea of paying for a fast didn't make much sense to me, but for some reason the idea of a seven day fast did. And so, for the next week I sat on the roof of my guest house where I partook of one very thin slice of papaya and drank two cups of water a day, as I sat reading Eckhart Tolle's classic mindfulness book The Power of Now. On the fourth day of my fast I heard that the Dalai Lama was 'in town' and was giving a talk that afternoon. To get the strength to walk the two and half hours up the mountain, I added a slice of apple to my diet (which didn't exactly give me that much more energy) and set off, water bottle in hand, to hear some words of wisdom from the person known as one of the world's greatest spiritual leaders, Tenzin Gyatso, the fourteenth incarnation of Avalokiteśvara, the Buddha of Compassion.

I arrived in the midst of a huge festival of around 3000 Tibetans; street performers, food vendors, traditional musicians and dancers, monks, animals and hundreds of kids colorfully clad in their best clothes running around excitedly waiting for their leader to appear. The Dalai Lama finally showed up and proceeded to give a forty-five minute speech, completely in Tibetan.

I wasn't amused. Lack of any real nourishment for four days plus exhaustion from climbing the mountain, coupled with the amazing local dishes that everyone was so generously offering me of which I could not partake, plus the total lack of any understandable inspiration from the man himself left me in a rather bad mood. I set off for home in less than high spirits and moved slowly through the crowds finally leaving the noise behind as I began my descent to Mcleod Gang.

And then it happened.

I was halfway down the forested mountain path as the sun was setting over the mountainous horizon, when out of the blue it hit me in the most stunning and clear way - The Power of Now...!

"See if you can catch yourself complaining, in either speech or thought, about a situation you find yourself in, what other people do or say, your surroundings, your life situation, even the weather. To complain is always non-acceptance of 'what is'. .... When you complain, you make yourself into a victim... The pain that you create now is always some form of non-acceptance, some form of unconscious resistance to 'what is.' "[60]

I stood there for a few awesome minutes as the sun shone gently on my face through the trees; feeling total presence and bliss as I had truly tapped into the power of accepting 'what is.' I almost floated the rest of the way back to my guesthouse and spent the next three days on a spiritual high, as the hustle and bustle of life in this busy haven of locals and spiritually seeking travelers unfolded around me. As I sat on the roof, water bottle in hand, watching the eagles soaring, silhouetted against the sunrise and sunset over the Kangra Valley, I enjoyed what actually felt like enlightenment, or at least the closest I had come yet.

After two months in McLeod Ganj and Bhagsu,[61] I decided it was time to move down the valley to Rishikesh, the Yoga capital of the world, considered one of the holiest cities in India. It is most famous in the West as the location of the Beatles' Ashram, where John, Paul, George

---

[60] Power of Now Chapter Four
[61] The next village two kilometers up the mountain

and Ringo spent time with their guru, the Yogi Maharishi Maharesh, in February 1968. While there, they composed around thirty songs including John Lennon's 'Happy Rishikesh Song.' The ashram itself is now deserted, overgrown and run down; a perfect playground for troops of Hanuman Langur monkeys, with their shiny silver coats, black faces framed with white fur and long wiry tails. These huge beasts rise up intimidatingly on their hind legs, reaching taller than a man and are not afraid to grab a bag of nuts or other food out of the hands of unsuspecting tourists. Rishikesh is spread out on both banks of the River Ganges, which at this point, close to its source high in the mountains, is usually crystal clear and ice cold. 'Mother' Ganges is considered a god by the Indians and many thousands of people ablute in her every morning and evening, even though by the time she reaches Varanasi, around 1000 kilometers eastwards across the plains of Northern India she becomes, to quote The Economist, [62] a "brown soup of excrement and industrial effluents" and is "not considered safe for bathing." During my time in Varanasi, I saw at least two dead cows and a dead human floating down the river, no doubt on their way to their watery resting place in the Indian Ocean, another 750 kilometers across the dry plains of Bihar State.

Downtown Rishikesh is like any other Indian city, crowded, noisy and polluted, not a place the average person would want to spend too much time. Yet just two kilometers up river it turns into a strikingly colorful spiritual wonderland. Two huge suspension bridges, Ram Jhula and Laxsman Jhula, built by the British, span the width of the Ganges, connecting the two banks which are both crammed with ashrams, temples, dhabas (restaurants), postcard and incense shops, spiritual book stores, travel agents and internet shops, all catering to the bustling market of souls seeking enlightenment and peace of mind. Bare-foot Sadhus (ascetics) dressed in light orange robes, Indian pilgrims from all

---

[62] "India and pollution: Up to their necks in it", The Economist, 27 July 2008.

parts of the sub-continent, polio ridden beggars and street kids, European and Israeli spiritual seekers and several cows and monkeys all bustle their way along the riverbanks and across the bridges, each looking for whatever will satisfy their particular hunger. Rishikesh is like a spiritual supermarket with an eclectic mix of courses on offer, yoga, tai chi, meditation, breathing workshops, massage, healing, reiki, Indian music, ayurvedic cooking and chanting to name but a few. After checking out the scene for a few days, I finally signed up for the intensive fifteen-day yoga retreat at Yoga Niketan up on a ridge with stunning views of the Ganges.

Yoga is one of the more well-known paths to Westerners, although the body positions (asanas) so popular in the West are only one step on Patanjali's eight fold yoga path. [63] Yoga means unification and is an attempt to unify body and mind, individual soul and universal G-d. There are many different philosophies, ideas and practices in India which have been conveniently lumped together by the West under the title 'Hinduism'. In reality, there is no one religion called Hinduism, and it's quite possible that several Hindus may each believe very different things. The basic philosophies are thought to be ancient, with the earliest known holy book, the Rg Veda, written in around 1200 BCE (approximately 112 years after the giving of the Torah at Sinai in 1312 BCE). There are six orthodox schools which believe in the authority of the Vedas, yoga being one of them, and several non-orthodox schools who don't.

I was in my element on the course, learning discipline and self-control, understanding and experiencing the connection between mind and body, individual soul and universal consciousness. I used my free time each

---

[63] The others are Yama – commitments, niyama – avoidances, prayama – breathing techniques, pratyahara – cutting out the sense perception and then three increasing levels of meditation – dharana, dyana and samadhi – the final level where individual soul merges with greater Divine soul

afternoon to take tabla[64] lessons on the banks of the Ganges, and every day at sunset I joined the aarti puja, a prayer service on the other side of the river with bhagans (devotional hymns) singing praises to the river, while floating candles down it to wash away sins and reflect on the day that had just passed. Aside from the physical postures and full body cleanses (I'll spare the details), I learnt basic yoga philosophy, including about the four main paths of yoga:

• Karmamarga - the path of action ensuring that all you do in the world conscious and for a higher purpose;

• Jnanamarga, the path of intellect, involving understanding the nature of reality and the path of enlightenment through studying the ancient texts and writings of the masters;

• Bhaktimarga, the path of devotion, involving surrender to and love of the Divine, through prayers and chanting and hymns; and finally

• Rajamarga, the path of meditation, the art of quieting and focusing the mind, connecting to the divine oneness permeating all of creation.

Many afternoons I would sit meditating next to the Ganges, a simple meal of rice and curried potato wrapped in a banana skin, tabla resting by my side, hardly any possessions or attachments to the physical world, no way to be contacted, no responsibilities.

I'd smile to myself, truly content, saturated with the feeling that I was tapping in to the consciousness of the masters, ascetics and sadhus, finally walking in the footsteps of the holy men of the past, following the path of the Yogi, uniting body, soul and Divine Soul in the ongoing dance of life.

---

[64] A classical Indian drum

'Renunciation! Renunciation! You must preach this above everything else. There will be no spiritual strength unless one renounces the world.
　　　　　　　　　　　　　　　　　　　　Swami Vivekananda.[65]

'A person will one day give reckoning for everything his eyes saw which, although permissible, he did not enjoy'
　　　　　　　　　　　　　　　　Jerusalem Talmud Kiddushin 4:12

The main Indian philosophical schools, called Vedanta, teach a very similar mystical world view to Judaism.[66] In brief, there is the Infinite Conscious Creator. All of creation is a manifestation of the Oneness of this Being; distinct yet not separate at all, (seeing as if anything was separate from Gd then Gd wouldn't be infinite). On one hand, the Divine is totally beyond the creation and unaffected by it, i.e. transcendent (the world is in G-d) this is why Hashem is often referred to as HaMakom – the Place, the space in which everything exists.[67] At the same time, the Divine is the 'material' of all creation and is totally within it, i.e. immanent, as the famous kabbalist Rav Moshe Cordovero said:

"*The essence of Divinity is found in every single thing — nothing but it exists. Do not attribute duality to Hashem. Realize, rather, that Hashem exists in everything. Do not say, "This is a stone and not Hashem." Rather, all existence is part of Hashem, and the stone is a thing pervaded by Divinity.*"[68]

---

[65] A key figure in the introduction of the Indian philosophies of Vedanta and Yoga to the Western world in around 1890
[66] There are actually a few different schools of Vedanta; Advaita, Vishtadvaita, Madhavadvaita to name a few, with some significant differences in philosophy, but for the purposes of this chapter I'm keeping it simple and general.
[67] You ever wondered what the universe is expanding into?
[68] Shiur Komah to Zohar

Quantum physics has now come to show that all of reality is made up of one unified 'substance,' quark, particle or wave. In fact, recently it has been suggested that the whole of 'physical' reality is actually really made of consciousness, something that Judaism has been teaching for thousands of years.[69]

The logical consequence of this is that our soul – called Neshama in Hebrew and Atman in Sanskrit - is a pure, manifest piece of Divinity referred to as 'Chelek Elokai Mima'al'[70] in Jewish sources and 'Tat Tvam Asi' in Sanskit, distinct but not separate from the whole. As Rabbi David Aaron says so poetically "You are someone - You are some of the One!" This philosophy is called Panentheism (not to be confused with pantheism which teaches just that the universe is G-d and G-d is the universe and nothing more).[71]

When I first arrived in yeshiva, the rabbis I shared the Eastern mystical philosophies with were very interested but not all that surprised. They showed me at the end of the Torah portion of Chayei Sara, that around 437 years before the writing of the Rg Veda, in 1637 BCE, Abraham sent six of his sons 'Eastwards to the East country,' with gifts. We learn that these were gifts of certain spiritual wisdom and ritual practices, which may have shaped the philosophy of the East. There are many examples which I have come across that support this idea, for example the name of the creator deity is Brahma, and he has a consort named Saraswati – strikingly similar to Abraham and Sarah. Legend has it that Saraswati was created when Brahma took a piece of his own body from which he fashioned her – much like Adam and Eve. These are just a

---

[69] See Prof Don Hoffman's work on YouTube
[70] See Tanya Part One Chapter Two
[71] Panentheism is the belief that the divine interpenetrates every part of the universe and extends, timelessly (and spacelessly) beyond it. Unlike pantheism, which holds that the divine and the universe are identical, in panentheism, the universe and the divine are not ontologically equivalent. God is viewed as the soul of the universe, the universal spirit present everywhere, in everything and everyone, at all times – Wikipedia

couple of many examples I've come across which may just be a coincidence, although we don't really believe in coincidences.[72]
Although parts of the core philosophies are similar, or at least were in the past, in practice the two religions are very different indeed.
Firstly, Hindu philosophies have been mixed with local animist and spiritualist religions, have personified the Divine energies and have created a pantheon of deities which people now worship and even attribute individual power to. This of course is antithetical to core Jewish beliefs and practices which teach that there is no power apart from Hashem and it is forbidden even to make, let alone worship, any statue or image. The other major way the two paths differ is in our relationship to the physical world. Hindu philosophy teaches that this world is Maya – an illusion. Not unlike the Buddhist idea of detachment, these philosophies in general expound that the way to achieve moksha – 'liberation,' is to renounce attachment to the physical world, freeing ourselves from the shackles of the body and ego which pull us down, control us and ultimately keep us from connecting to our Divine essence. Like Buddhist monks, Hindu sannyasins don't get married, they refrain from physical intimacy, they generally eat as little as is needed to sustain the body, renounce all worldly pleasures, have no money and very few possessions and live ascetic lifestyles, free from the stresses of everyday life and able to concentrate on meditation, yoga and other spiritual practices.[73]
Whereas Eastern philosophies in general teach us that to be spiritual we have to detach from the world, and the West tends towards materialism

---

[72] In fact the Hebrew letters of the word for coincidence – מקרה – can be re-arranged to spell רק מ'ה – only from Hashem, ה רקם – G-d embroiders, קרה מ – happened from Hashem and of course קרמה – karma the law of Divine justice as taught in 'Hindu' philosophy

[73] There are some esoteric schools that are more focused on the physical world and the path of the householder - for example Kashmiri Vaishnavism, however, in general, the ideal is renunciation, as seen by the opening quote from the highly respected guru Vivekananda

and physicality, Judaism teaches that we need to balance and integrate the physical and the spiritual aspects of ourselves – our body and our soul. [74] Our sages teach that rather than being a physical being (a body) that is trying to have a spiritual experience and connect to our soul, we are in fact spiritual beings (soul) that have a body, the aim of which is to use physicality as a tool to revealing our spirituality.

At first glance, as is very often the case, it seems that Jewish sources actually teach two opposing ideas in this area:

• On the one hand, Jewish sources teach that Hashem wants us to partake fully in and enjoy the physical pleasures of the world. The first two commandments given in the Torah are 'Pru uRevu,' be fruitful and multiply and 'eat of all the fruit in the garden.' Our festivals are accompanied by lavish meals, we are meant to have special clothes for special occasions and Purim takes indulgence in physicality to the next level. The above quote from Jerusalem Talmud even suggests that at the end of our time here we'll have to answer for not enjoying the physical pleasures of the world. So we see that Judaism teaches that Hashem, like any loving parent, wants His/Her kids to enjoy some ice-cream once in a while.

• Yet other sources teach us that'

'The way of Torah is to eat bread with salt, drink a little water and sleep on the floor'[75] and Rebbi[76] claimed that he got no pleasure from the world, even in his little finger! The Talmud[77] teaches that we should sanctify ourselves by abstaining also from that which is permitted to us

---

[74] Maybe this is why Israel is in the middle, in between the East and West …
[75] Avot 6:4
[76] Rabbi Yehudah HaNasi, redactor of the Mishna, quoted in Ketubot 104a
[77] Yevamos 20a

and the Rambam teaches that we should only eat what we need to maintain our health.[78]

The Zohar even calls this world alma d'shikra, a world of falsehood. From this it is clear that self centered pleasure taking is not the way either.

Of course, these two ideas are not contradictions and taken together, they demonstrate the true path. Judaism teaches that the ultimate spiritual path is not to shun the world and remove ourselves from it, nor to get lost in materialistic hedonistic pursuits. Rather, in order to be on a truly high spiritual level we need to be very fully in this world, just in such a tremendously mindful way that we actually use physicality as a tool to become connected to the greatest spiritual heights.

How is this achieved? Let's take eating as an example. We love eating and a major part of almost every festival and holy day is the seudat mitzvah – the mitzvah meal. However, before we put any food in our mouth, we make sure it's kosher, we say a specific blessing, eat consciously (or at least are meant to), share words of Torah during the meal, say a specific blessing at the end – in this way we are turning eating itself into a spiritual practice. The same goes for all other permitted physical pleasures in the world. Monks don't have physical intimacy, in the West it's a free-for-all, Judaism says physical intimacy with the right person, at the right time, for the right reasons, as an act of giving and connection is a very high meditation between souls. In the East - no money, in the West - mo money, Judaism says you can have money just use it for doing good in the world. Physical pleasure for its own sake is animalistic, temporary and disconnects us from spirituality. Pleasure as the reward for or consequence of yearning to uplift the world and bring G-d consciousness into all mundane activities is the greatest, most authentic physical pleasure there is.

---

[78] See Rambam Hilchot Deot 3:2

The question is one of motivation and goal. If we are pataking just for selfish temporary pleasure we are no better than animals, yet if we are partaking in order to build a conscious connection to our higher Self, others and Hashem, then we are higher than angels.

This is really what it means to be holy.

People often ask me "I'm a good person. I have good traits; I'm loving, kind, generous, honest, I do good things in the world. Why do I need the Torah?" It's an excellent question with a simple answer. The goal of the Torah is not to be 'good.' I don't need Gd to come out of the heavens to tell me to be good! It is just basic common sense, and there are plenty of good atheists walking around. Being good is the most basic level. Our sages teach us that Derech Eretz Kadma L'Torah - Being a decent human being comes as a prerequisite to keeping Torah. The actual goal of the Torah and mitzvot is to help take us to the next level; being Holy. In the Torah Hashem urges us;[79]

"Be holy because I am holy." So, what does it mean to be holy? Although the most common understanding of Kadosh implies separation and sanctity,[80] probably the most famous verse in which we find the word kadosh says *"Holy Holy Holy, Hashem master of Legions, the whole world is filled with His glory!"*[81]

That doesn't sound separate to me. In fact, it sounds totally present and immanent. This suggests that to be holy means to have the balance between being fully present and involved in the world while also transcending it in the sense that we are not controlled or defined by it and we interact with it in a conscious and soulful way. As the verse says, just like Hashem is holy – ie totally beyond the world (transcendent) yet totally within and involved in it (immanent), so too we must be the same.

---

[79] Vayikra 11.44, 19:2
[80] Rashi: Vayikra 19:2
[81] Yeshayahu 6:3 - which we say as the pinnacle of the daily morning and afternoon prayer

This, however, is no easy task and it's because of the pitfalls of trying to walk this path that many religious philosophies stress renunciation of the physical. It just seems too hard to be involved in the world without getting sucked in! Human nature is to be pulled towards physical desires such as food and ego desires such as honor - therefore many philosophies rather just withdraw completely.

The truth is that without some sort of guidance and system it *would* be almost impossible to achieve this. It's for that reason that Hashem had to give us the Guidebook, the Torah, an instruction manual which teaches us how to tread the fine line between physicality and spirituality. Far from being burdensome obligations that Gd has imposed on us, mitzvahs are in fact physical acts that have been given to us as connection points, through which, if done with intention and joy, we can reach the highest level of spirituality and truly connect to ourselves and our purpose; an intimate relationship with Hashem. [82]

There's a story about a Buddhist monk, a college student and a Rabbi who enter a room with wine bottles free for the taking: The college student walks into the room, starts drinking, singing with delight until he finally starts crying about his life, throws up and passes out. The Buddhist monk walks in and noticing the effect of alcohol undertakes never to even touch a drop and walks out. Then the rabbi enters. He sees the wine on the table, checks a bottle and sees that it is kosher, opens it up and pours himself a glass. He takes a few moments to say a blessing, showing gratitude, being present and connecting to Hashem, sanctifying the wine. Then he has a few sips and takes out the Torah book he has been learning and sees some ideas that he hadn't grasped before. He has another glass, feels good, sings a few uplifting melodies before making an after bracha, cleaning up and leaving to go and give some charity.

---

[82] The word mitzvah itself is related to the root Tzavta – meaning a connection.

In moderation and with consciousness, the physical world can be enjoyed and sanctified and ultimately used to serve our spiritual aspirations.

Like all spiritual growth, it's important to take small steps in this area. See Appendix E for some exercises and practices you can implement in order to integrate these ideas into your daily life.

After spending so much time in meditation retreats, yoga courses, fasting and sitting in caves and temples following my breath going in and out, I started to get the feeling that sitting in a beautiful place, separated from the problems of the world, trying to overcome my ego while so many people were suffering around me, was, well, a little egotistical. It seemed clear to me that the world can't have been created for us to overindulge and abuse it, but it also didn't make too much sense to me that the actual goal was to completely renounce it. There had to be a balance. At the time, the balance that spoke to me most was to live a simple spiritual life of meditation and peace in the Indian countryside, while running an orphanage to help the lives of as many vulnerable kids as possible. Little did I know that, even though it was a truly noble plan, so much more than that would soon be asked of me.

If I had more space I'd finish this chapter telling you about the two months I spent in Bodhgaya, the town dominated by the temple built next to the Bodhi Tree where Buddha was enlightened; where I volunteered in Samanwaya Orphanage run by Dwarko Sudrani, a fascinating man not least of all because he was Gandhi's last living disciple. I'd share some of the conversations I had late into the night with Suresh-ji, who had devoted his life to caring for the thirty kids in his orphanage in Baghar including the time the Naxalite-Maoist freedom fighters who were terrorizing the state came to him to seek advice in the middle of one of our chats about the benefits of yoga for the kids. Once they left Suresh-ji told me how relieved he was they hadn't kidnapped me. I'd tell you about my time down south in the ashram of one of India's most famous gurus, where I happened to be when the huge tsunami struck India, Sri Lanka and Thailand - which I only found out about four days later when I checked my e-mails and saw one from my father to all the heads of the Indian states asking for any news of my whereabouts. I'd tell you about my trip to Manali when I got stranded on a mountain with a dreadlocked sadhu who suddenly became paranoid because he ran out of hash and wanted to rush home in the middle of the night, meaning I had to charge down the ragged rocks behind him with only a makeshift torch made of his shirt wrapped around a branch dipped in some highly flammable substance and set alight to show us the way. I'd tell you about the burning bodies on the banks of the Ganges in Varanasi, my near-death experience in the thick fog that descended suddenly on to Triund Mountain, my visit to a local Indian hospital in Pondicherry where the injection made me vomit and my whole arm swell up like a tennis ball as tens of bleeding and moaning Indians were lying on the floor all around me. I'd fill you in on 35-hour train journeys, high altitude hikes to swim in lakes, rivers and waterfalls and the time I almost got stuck up a mountain for two months as a freak blizzard blocked of all roads and footpaths leading back to town.

Yet, as I said at the beginning, India is too vast to be squeezed into one chapter of a book, even the longest one. Suffice it to say, it taught me in so many ways the lesson of accepting what is, as it is. In India you can't fight against nor run away from the reality, the heart wrenching poverty, the breath-taking beauty.

It was only much later that I discovered there could be an even higher level of enlightenment, a higher level of acceptance of 'what is'.

The Talmud[83] teaches us that not only should we accept 'what is' because that's the reality, but that everything that happens is actually for our good, specifically orchestrated by a Loving Guide in order to help us grow and reach our potential and purpose in life. The key is not to just accept 'what is,' but to understand fully and feel deeply that 'what is' is absolutely the best possible thing for us, however difficult it may seem on the surface. In this way, we not only avoid suffering from it, we learn to embrace it and use it as our tool to achieve insight and growth. To know this and live it fully is the real enlightenment, the real power – The Power of Emuna (Trust ... that everything is perfectly orchestrated by a Loving Parent for our best).

After a few months in yeshiva, I was sitting on the roof of my dorm room thinking back fondly to my time on the banks of the Ganges. I started to wonder if I had given up the real quest for holiness and spirituality and settled for something less. I had always felt Judaism was an outdated religion, void of spirituality and truth - old man in the sky telling me what to do or he'd punish me, full of can't dos ... Had I lost touch with the ideological adventure seeker who wouldn't settle for less than the truth and the most peak experiences? Had I let go of that inspired inquisitive part of me which craved connection, ultimate

---

[83] Brachot 60a

knowledge, the mystical experience? Had I given up my 'Self' to do everything the rabbis, ancient and modern, tell me to do? It took just a cursory look to assure me that quite the opposite was true. The Judaism I encountered revealed to me that every moment is a meaningful new opportunity to connect joyfully and powerfully to the magic of life.

I realized that the path I have chosen is a balanced synthesis, if not the basis, of all the yoga paths I learnt about many years earlier. Conscious action (karma marga) as laid out in the mitzvahs, guiding us how to change ourselves and the world around us through giving charity, relating to our fellows in an honest and wholesome way, using the physical world as a tool to be spiritual. Intellect (jnana marga) pushed to the limit by Torah study, sharpening the mind and perception of the world to an incredible degree by gaining insight into the deep teachings of real masters. Yearning devotion (bhakti marga) in three prayer services and over a hundred blessings a day, powerful heart moving niggunim (songs) in candle lit rooms as Shabbos is going out, personal prayer in forests and fields and finally deep meditation (raja marga) as practiced by the masters of Kabbalah. Far from being the dull ritualistic religion I had thought it to be, Judaism turned out to be the comprehensive path to consciousness, understanding and self-realization, perfecting the self and the world around us, unifying our individual Soul with the one great Divine Soul. In short, I discovered that Judaism is the Ultimate Yoga.[84]

---

[84] Yoga means unification, ie realizing the innate connection between the neshama and Hashem

Photos

Annapurna Base Camp: Nepal. p88

Teaching Buddhist Monks: Laos. p39

Photos

Shaolin Training Fa Wang Si Temple: China. p77

Aikido Training: Japan. p96

Photos

Baghar Orphanage, Bihar: India. p72

Shikoku Island 88 Temple Pilgrimage: Japan. p102

Photos

Yeshiva in Jerusalem: Israel. p112

Walking the Israel National Trail: Israel. p117

Chapter Seven
China
Shaolin Masters and Survival Missions

*J*ames and I set off into the wilderness of ChangBaiShan, the mountain range which forms the natural border between China and North Korea, with just a 5 x 2 meter sheet of plastic, four kilograms of rice, some noodles, two kilograms of kimchi, a packet of cookies, a couple of cigarettes and a large machete which we bought at the Erdao Baihe Central Bus Station. Although we both grew up in England, our childhood couldn't have been more different. A few years younger than me, standing over six-foot tall, with a long mop of ginger hair and a scraggly beard, James grew up in the English countryside, working the land with his hands, hiking the windswept mountains of Scotland, going to church on Sunday - a far cry from my middle class North London Jewish upbringing. Yet we hit it off the first time we met, as we bumped into each other both wandering round the quaint former British hill station of Shimla in Northern India, looking for a place to watch the rugby world cup final between England and Australia. After witnessing England's historic 20 - 17 victory on a small fuzzy TV in a hotel lobby, we spent a couple of weeks travelling together, playing chess, hiking in the mountains and volunteering in the orphanage in Bodhgaya.[85] We eventually went our separate ways but stayed in touch, and when the chance for adventure arose in China, I knew he was the man to partner

---
[85] See Chapter 6

with. We had agreed that under no circumstances would we return to civilization in under six days, even though we heard that a few weeks earlier a foreigner had wandered off into the mountains and crossed into North Korea by mistake, where he was shot, or just arrested, depending on which version of the story you chose to believe. Either way, it wasn't something that was on my bucket list. We entered the national park and as soon as we were out of sight of the rangers, ducked off the main trail and ran through the undergrowth until we came to a small stream running down the mountainside, a runoff from the sacred Tianchi Lake which filled the center of the now extinct volcanic crater. We crossed the stream, walked a bit further down into the valley until we found a perfect spot and set about using our plastic and some wood to form a makeshift home. We created an A-frame shelter with two branches forming an 'A' at the one end and a long cross beam which we lay across the top and lodged into a tree at the other end. We draped the plastic over the structure and pinned it down on one side with some sharp wooden pegs we had carved with our machete, leaving the other side wrapped around a branch and unpegged so it could be lifted up and closed as desired - our own convertible tent. Then came the business of gathering as much grass as we could to cover the floor of our home, creating a comfortable natural mattress, while at the same time collecting wood for our fire and branches and leaves to camouflage the structure from the prying eyes of park rangers and North Korean snipers (okay so our imagination ran away with us a bit, but it was all part of the fun). As we set up camp, we could see the path winding its way up the crest of the mountain, with unsuspecting tourists coming back from the crater, as we nestled in for the night.

And then it began. Slowly at first, then gradually building up to a consistent and monotonous downpour as the north-east China rains swept across the mountain and continued almost non-stop for the next three days. That meant no fire, which meant no heat or cooked food *leaving us to survive on one cookie and a couple of spoons of kimchi a day. It also meant very little exploring the area outside of short forays when the rain let up for a few minutes - seeing as we only had the clothes on our backs and no way to dry them.*

We were pretty cold and hungry and didn't have much to do as we crouched in our meter tall, five-meter long enforced prison for a few days. Yet, unlike the weather, we were very far from downcast! This was it! We were living the dream, albeit not 100% as planned, but the dream nonetheless! Our friends were working in office jobs and here we were, in the middle of the mountains, surviving in the wilderness, being creative, minimal rations, totally free and wild, two English boys pretending to be survival experts, us against the elements, maybe physically confined, but emotionally soaring over the mountains of one of the most mystical and stunning places on earth.

I arrived in China focused on pursuing my dream of training martial arts in the mountains that had seen centuries of enigmatic Taoist sages, real and mythical martial arts masters, and the rise and fall of several great Emperors and their dynasties. Shaolin, in Henan Province, is the world headquarters of Kung Fu. In the 5th or 6th century Bodhidharma, an Indian monk, is said to have come to China whereupon seeing that the monks were being constantly robbed by groups of bandits, decided to teach them how to defend themselves. Legend has it that he was meditating, gazing at a wall for nine years, when in the seventh year he kept falling asleep, so he cut off his eyelids and tossed them away over his shoulder. Where they landed grew the first green tea plant, leaves in the shape of eyelids, which produce tea which keeps you awake in a healthier way than having your eyelids removed.

Enlightened Jewish Master — China

I arrived in Deng Feng, a town about fifteen kilometers from the Shaolin Temple, at the foot of the holy Mount Song, where tens of martial arts schools cater to thousands of practitioners of many different fighting styles. I started asking around for the best schools and began hearing rumors of a picture perfect temple up in the mountains which had a small kung fu academy in it. I expectantly made my way up the winding mountain roads for around two hours, when behold, rising out of the mist I got my first glimpse of Fa Wang Si, the second oldest Buddhist temple in China. Stretching up level after level into the Yuzhu Peak of Mt Song, Fa Wang Si, apart from housing a small martial arts academy, attracts thousands of tourists and pilgrims each year. On each level is a temple building with ornate banisters lining the stone staircases. Statues of fearsome dragons, lions and demons protected the deities in the temples, large rocks made great meditation seats and small lawns and sculpted bushes and trees added splashes of natural green to the Hollywood movie like setting.

Training began at five am, when the bell would sound to wake us from our all too short slumber, certainly not enough sleep to help our bruised and aching bodies recover from the grueling training from the day before. We began the new day with some stretching before heading down to the bottom of the mountain from where we'd begin our run back up to the temple. On the first day, I hardly made it a quarter of the way up before my lungs felt like they were going to burst and my legs refused to go any further. By the time I left five weeks later I could comfortably make it all the way and had cut my time by three quarters. Then we had breakfast which, like lunch and dinner, consisted of very oily rice, fried eggs and some stir fried vegetables. The rest of the day was spent stretching, practicing cartwheels and flips, perfecting the five main animal forms - tiger, dragon, snake, crane and leopard - weapons training (my favorite was the guun or long staff) weight training, tai chi and personal meditation. The master spoke almost no English and my

Mandarin never really got past the basics, yet he always managed to get his point across, often with a firm blow of his bamboo stick to the back of the thigh. When we had some free time, we went exploring the mountainsides, found some small lakes to swim in and at night we'd sometimes go into town for a beer and karaoke. But usually we were too worn out from the day's training to have the energy to make it down to town and after dinner and a final training session we collapsed into our beds in the cramped dormitory, exhausted yet exhilarated from another day of pushing our bodies and minds to the limit, conquering physical and mental obstacles, living up to the Chinese name my master gave me - 和平战士 Heping Janshe; Peaceful Warrior.

"Don't pray for an easy life, pray for the strength to endure a difficult one."

Bruce Lee

"According to the effort is the reward." –

Ethics of the Fathers, 5:23

In all walks of life, from martial arts to music, personal growth to business, we see that to achieve anything takes focus, discipline, personal sacrifice and effort. In fact, a comfortable life - although fine - is never going to be extraordinary and memorable.

Take a sports person, for example. They watch what they eat, don't go out drinking with friends, get up early, do weights, stamina training, day in and day out pushing themself to the limit to achieve their goal. That is the only way to success. My least favorite phrase in the English language is 'It's easier said than done.' Easier said than done. I don't dislike it because it's not true - it *is* true. It is just that the phrase is useless and defeatist, just a statement of fact, usually an excuse for not getting things done. Once a guy in yeshiva asked me how he could lose some weight. I suggested going to the gym and focusing on cardiovascular workouts, doing high reps of lighter weights, cutting sugars and fats out of his diet. When he replied "Wow Rabbi, that's easier said than done!", I gave him a bit of a confused look and informed him that while indeed it is easier said than done, just saying it wouldn't help him lose any weight, whereas doing it most likely would. Life, especially a great one, is not easy and comfortable, although many of us would like it to be. All achievement comes from overcoming some sort of struggle. Rav Noach Weinberg taught that people mistakenly think that pain is the opposite of pleasure. However, the truth is that the opposite of pain is no pain, or comfort. Pain is often the price we pay to achieve true pleasure, as they say 'All growth comes out of the comfort zone." When we truly believe that something will give us pleasure, we are willing to put in the work, take the pain to achieve it, whether it's working hard for exams or going to the gym so we look good.

At some point in life, just like the sportsperson, people need to ask themselves what they value, what they are passionate about, what they are living for, and therefore what they are prepared to put effort into:

Do you want a healthy relationship with your spouse? That takes effort.

Do you want to be rich? That takes effort.

Do you want to raise well balanced and happy kids? That takes effort.

Do you want to become a great person? That takes effort.

You want to help people, find a fulfilling job, learn a language or instrument, climb mountains – anything really worthwhile takes effort.

Even people who claim to be a bit lazy, once they have a goal they value are willing to put in the effort to attain it. I once asked a group of college kids if they found it hard to get up in the morning. They all put up their hands. 'What if I offered you $50,000 to meet me at four am and do ten star-jumps, who would be there?" All hands went up again.

So I pointed out that really none of them find it hard to get up, it's just that they don't value what they're getting up for.

The foundation of all discipline, and therefore what takes the greatest effort, is self-control. This means to be able to do, say and even think what we know is right and healthy even when it's difficult and we don't feel like it.

The Torah exhorts us to develop self-control when it tells us not to chase after our hearts and eyes,[86] just going along with however we feel and whatever we desire. Humans (unlike animals who - unless there is an external factor - can't control themselves) don't have to eat just because we feel hungry, don't have to shout at our kids just because we're in a bad mood and can still go to the gym, even when we feel tired. Someone with self-control can hold themselves back from doing detrimental things and push themselves forward towards greatness, not always at the whim of their thoughts and desires.

The master of self-control, and the only person in the Torah who we refer to as 'HaTzaddik',[87] was Yosef. The normal reason given for this is because he managed to overcome his physical desires in the face of constant seduction by his master's wife. This in itself takes a large amount of self-control, overcoming the most powerful drive we have in order to do the right thing (especially seeing as he was a good looking 17 year old far from home!).

Yet there was another occasion where he demonstrated this trait to an even greater degree. As a young boy, he shared with his brothers that he had a dream in which they all bowed down to him. This really upset them and they ended up selling him into slavery. He went down to Egypt where after several trials he rose up to become the second in command of the whole country. Many years later, there was a famine in

---

[86] Bamidbar 15:39
[87] Although there were many tzaddikim, he's the only one we call HaTzaddik

Israel and the brothers were forced to go down to Egypt to seek food to bring back to Yakov their father. When they arrived in the palace they bowed down to the viceroy of Egypt, who was, unbeknownst to them, Yosef, ultimately fulfilling the dream from all those years ago. At that moment, Yosef had the best opportunity of all time to say perhaps the four most satisfying words known to mankind - I told you so! It's an awesome feeling to be right, to have your ego boosted, to be the victor, especially after being betrayed. Yet Yosef held himself back from needing to feed his ego, from needing to be right. Instead he only had their best in mind and didn't even blame them for their treachery. This is the sign of a true master, someone who has the discipline and self-control to do what is right, physically, emotionally and spiritually in any situation.

On one level, and the way most people look at it, self-control means the ability to control one's thoughts and desires, to not just go with them and let them control us. However, that implies that the thoughts and desires are my Self that I need to control. From a Torah viewpoint, self-control means that I am not controlled by anything external to me - not by what others think of me, not by the weather, how well my team are doing, my financial situation, and not even by my own thoughts and emotions. It means that my Self, i.e my soul, my true essence, is in control, making the decisions, overriding the thoughts and desires of my body in order to lead the way towards becoming a tzaddik. All we need to do is have the presence of mind in any situation to ask ourselves 'What would the best me, the soul me, think, say, do in this situation?' and act accordingly. To the extent we manage to do this and act according to what our best selves dictate, we will reach success and true mastery in life.

It takes a lot of discipline to live the path of an Enlightened Jewish Master. It's not easy to get up to pray every morning, to say one hundred blessings a day, to hold back from eating certain foods, looking at and listening to certain things and even thinking certain thoughts. For a Jewish businessperson to give up a day of work a week, or a pregnant woman to pass up on some tasty Ben and Jerry's after eating meat is not a simple matter.[88] Committing to do these things takes a strong will and value system and requires consistency and strength of character. Yet as we already learnt, if you value it enough you will put in the effort and reap the rewards. See Appendix F for some exercises and practices you can implement in order to integrate these ideas into your daily life.

It is certainly a high level just to have the self-control and commitment to do these things every day; yet the goal of Jewish practice is not just about getting them done, it's about how we are doing them. The sign of a true Jewish master is that s/he is following the path not out of fear, guilt or habit, but rather with true understanding, consciousness, intention and joy.

If Fa Wang Si taught me one thing, it's that we are not going to achieve much while staying in our comfort zone. I still smile (and wince) as I remember my master's grinning face as he proudly shouted his only English phrase – "No pain, no gain, no pain, no gain!" before swinging, with relish, the bamboo stick towards the back of my already red raw thigh.

On the fourth day James and I woke up to silence. No rain pattering above our heads, no wind rustling the plastic walls of the tent. The sky was a clear blue and the sun warmed our faces as we emerged from our

---

[88] According to our mystical masters we don't eat milk together with or for a while after meat seeing as meat is an expression of death and milk gives life.

makeshift home. We stacked the wood up in a trellis format so the sun could dry it out with the help of the soft breeze that was now wafting over the mountainside. We washed off in the river, surveyed our surroundings and breathed the fresh air of the calm after the storm.

We were excited to get a fire going and cook up our rice and noodles but were a little concerned that the smoke may be seen from the trail. We gathered up our wood, cooking utensils and the food and headed out into the next valley where we nestled down and eventually got the fire started even though the wood was still pretty damp. Twenty minutes later, we were sitting down to our first warm cooked meal in a few days, and I can tell you, plain rice and noodles with a touch of salt have never tasted so good! After our little feast we quickly headed back to base camp seeing as at that point we weren't sure whether or not we had crossed into North Korean territory. We were so energized by our victory that we decided the time had come to make our way up the mountain to get the coveted view of the famous Tianchi Crater Lake. We were so energized from being cooped up in our tent for three days that we ran up to the summit where we were greeted with our first view of the 'Heavenly Lake.' We sat in awed silence for several minutes, taking in the view, breathing deeply and reveling in our achievement. It was another great example of Ben Hei Hei's wisdom from Pirkei Avot – 'According to the effort, so is the reward,' the more effort something takes, the more rewarding the result is. As the sun began to set we made our way back down to our camp site, and arrived just in time, as the clouds started moving in and the rain showers spread their way across the mountain once more. The next morning, we packed up our stuff and headed out towards civilization. After treating ourselves to our first hearty meal for a few days, we made our way back to Beijing from where we embarked on our next mega adventure – the six-day Trans-Siberian Express train journey which cut briefly through Mongolia and then across the whole of Russia to its final destination in

Moscow. From there, I travelled overland through Latvia, Estonia and Holland, before heading back to England to see my family. I ended up staying and working in London for three months, reconnecting with friends, family and Western civilization. Yet all that time I was planning my next adventure, to fulfil my lifelong dream of visiting the 'Crown of the World' – the Nepalese Himalayas.

Chapter Eight
Nepal
Himalayas at Last

Since the age of eight when I first laid my hands on the Willard Price Adventure books, I have been passionately enraptured by stories of survival and adventure. I list as some of my favorite reads 'Into the wild' and 'Into Thin Air' by Jon Krakauer which tell of true survival adventures with tragic outcomes and 'The Long Walk' by Slavomir Rawicz which details the 'true story' of a Polish prisoner of war who escaped from a Russian camp in Siberia and made an epic 6,500 km journey on foot through the Gobi Desert, Tibet and the Himalayas until finally reaching India in the winter of 1942. It was only much later that I found out that the story probably isn't true (still worth reading though), but it was certainly enough to light a fire for adventure in my heart. The Himalayas in particular have always fascinated me, mystically, geologically and aesthetically and I knew from the beginning of my journey that I'd be spending a decent amount of time in the Abode of Snow. At the same time, and looking back I'm not sure exactly why, I've never had a strong desire to do anything too crazy myself. Although the 1200 km solo hike around a Japanese island was pretty challenging, [89] it was never really life threatening and I was always within a few hours walk of civilization. The 1125 km forty day hike of the Israel

---

[89] See Chapter 9

National Trail[90] including ten days walking through the desert and a midnight encounter with a wolf, had its challenges but the trail in general was full of Israeli home hospitality and national park services. My near-death experience as the fog suddenly descended on the Triund Mountain in Northern India as I was making my way down could have ended pretty badly but it was only touch and go for a few moments and I probably could have survived one pretty miserable freezing cold night with some broken bones until someone would have found me in the morning. The six days in Changbaishan[91] had its challenges but ultimately was controlled enough to preclude any real disaster. Summiting Everest was never a dream of mine and solo expeditions to the North Pole never even crossed the radar. I'd like to imagine that it is because I felt full of wonder and alive enough to not have to risk my life doing crazy things, or maybe the adventures and travels I did have fed my thirst for nature, or maybe because I'm just too much of a mommy's boy to risk leaving her bereaved. Whatever the reason, I managed to get a taste of the wilderness and everything it has to offer without actually placing myself in too much imminent danger.

In any case, after a few months in England, I packed my bag once again, bid farewell to my family, who by now were getting used to saying goodbye (and planning their trip out to visit me), and headed out across the world, determined to dig deeper in my search for personal mastery as my warrior's journey was finally taking me to the mountains of my childhood dreams.

The Himalayan mountain range is a literally awe-inspiring expanse of snow covered mountains which cover 80% of Nepal's land mass. It is

---

[90] See Chapter 10
[91] Mountain on the border of China and North Korea - See Chapter 7

home to eight of the world's ten highest mountains including Mt Everest, whose peak at 8,848 meters makes it the highest place on earth. Tales abound of great sages and yogis hidden in the caves and forests, and many of the mountains themselves are considered sacred by the locals and adherents to the regional religions. Predominantly a Hindu country, Nepal also has a strong current of Buddhism running through all aspects of society, not least of all because Siddartha Gautama, the Buddha himself, was born there. The capital, Kathmandu, is a thriving, sprawling metropolis of over 1.4 million people, and while there are more modern buildings cropping up all the time, the colorful skyline is still generally taken up with traditional wooden and mud housing and stores, including temples on almost every corner and street markets selling all kinds of exotic food and other necessities. The conspicuous five colored prayer flags which pop up all over Northern India, Nepal and Tibet flap in the wind and the hustle and bustle of merchants and tourists making their way through the narrow streets of this ancient city, along with cows and chickens roaming free, means it's easy to imagine what it must have been like a five hundred years ago, as a hodgepodge of cultures and traditions converged on this important trade center on the merchant routes between India, Tibet and China.

Upon arrival, I immediately settled back into my backpacker mind set, no pressures, no worries, just curious and free to wander wherever I chose, the sparkle of life on the trail back in my eyes. After two days of acclimatizing, exploring and learning some Nepali I set off on the seven-hour bus journey to Pokhara. The fact that it takes seven hours to travel the 210kms between the cities gives a good impression of what the terrain and roads are like in this part of the world. Set on a large lake, framed by the awesome Annapurna Massif, Pokhara is a picture perfect gateway to some of the world's most stunning yet manageable treks, totally geared to the backpacker and trekking scene, which makes it Nepal's most popular tourist destination and a great place to experience mountain life.

Having settled into a guest house on the shores of the lake, two Americans I met on the bus and I booked a ten-day hike up to Annapurna Base Camp, a route that took us up through barren desert-like landscapes, beyond that through the forested lower slopes finally breaking through to the postcard perfect vistas of the Himalayas that I'd always dreamed about. Annapurna itself is the most dangerous of the world's fourteen over 8000 meter peaks, with 34% of summit attempters not making it back down. We spent our days walking through charming villages, farm animals crossing our path (or more accurately us crossing theirs), lush greenery all around, quaint cobbled streets and colorful mud houses, kids in their school uniforms playing football with rolled up clothes as a ball, ladders made of tree trunks with grooves cut up their length and the constant smell of burning wood, used to cook food and heat water for bucket showers. In the evenings the foreign trekkers would gather in guesthouses, exchange their hiking boots (or trainers in my case) for sandals to give the blisters some rest from rubbing, cold beers or warm teas in hand, sharing stories of their travels, playing cards and 'feasting' on dahl baht, the staple Nepali food of lentils and rice, or pot noodles and western chocolate bars that had been schlepped up from Pokhara to bring some 'luxury' to the trail. The attitude of the locals and hence most of the trekkers toward the mountains was one of utmost reverence and respect. Arrogance was out of place here, no-one 'conquered' the mountains; they were here long before us and will be around long after we depart from our temporary abode.

One evening around four days into the trek, with still another full day of hiking to reach base camp at 4130 meters, I was sitting on the guest house roof in the middle of the night, breathing the crisp thin air, the silhouette of Machupuchare[92] sharply lined against the dark blue sky, closer to the stars than I had ever been while still on the earth, the only noise coming from a gentle breeze whisping down the valley. I smiled as I contemplated the fact that just one week ago I couldn't have been further, physically or mentally from where I was now. Traffic and TVs, coffee shops and clubs, celebrity gossip, creature comforts, designer clothes, fast paced work life, sports mania, with nearly six thousand people per square kilometer, all running around on the treadmill of life. And here I was, sitting thousands of meters above the clamor of the 'real world,' no way to be contacted or to contact anyone, no technology, a small bag with a few essentials in it, nowhere to go, nothing to do apart from revel in the magical wonder of having finally made it into the mountains that had been enticing me all these years, totally serene, alone and present, feeling, in many ways, on top of the world.

"Hashem said to Moses, "Come up to Me to the mountain, and just Be there."

Exodus 24:12

The term Mindfulness is trending now days with mindfulness books selling millions of copies, yoga centers popping up left and right, meditation apps flooding the market, all turning the whole "show" into a multibillion dollar industry. Of course, this is hardly surprising given the opposite trend that it's coming as a reaction to. According to a recent survey, the average American spends 90% of their waking hours looking at a screen! Most groups I speak to admit that the first thing

---

[92] Fishtail Mountain, considered so sacred by the locals that no-one has ever been allowed to attempt to summit it, and it is now off limits to all climbers

they do when they wake up and last thing they do before they sleep is check their phones for messages and updates. Many people work in front of a computer, in the bathroom they are checking their smart phone, in the gym there are screens, at home is TV, social media and Zoom on the laptop, and many go to the movies or sports bar on their night off. One survey I read reported that people compulsively check e-mail seventy-four times a day and look at their phone over one hundred and fifty times a day. The average person has at least five notifications channels – Whatsapp, SMS, Facebook, phone and e-mail - buzzing and beeping all the time. Add to that Linkedin, Snapchat, Instagram, Tik Tok, Twitter, YouTube – we're being constantly bombarded by things vying for our attention. What this means is that we are very rarely, if ever, actually alone. In fact, the thought of having no distractions and just being alone with themselves really freaks most people out. When I tell groups that I spent ten days in silence with no reading, writing, phone or TV they can't believe it or understand how that is even possible. People feel the need to escape the pressures of life and the endless chatter of the mind and when not at work, many people spend their free time actively trying to escape rather than just sitting quietly and being comfortable with their own company. TV, video games and different forms of social media browsing make up four of the top five past times in America today. Some of the highest paid people in society are the entertainers; sports, music and movie stars, people who help others tune out of their lives and take a break from their minds for a little while. The problem with this method is that when the temporary break from life and mind chatter comes to end, when the final whistle blows, the drugs wear off or the credits start rolling, we're just back where we started. This can all lead to a rather meaningless and superficial relationship to ourselves, others and life in general, as people look to get the best selfie to share on Instagram rather than actually just being present and enjoying the moment. [93] Depth in life can only come from being truly present and engaged and meaning in life can only come from being present and engaged in meaningful things, that is, things that are actually leading us towards the goal and purpose of life - personal growth and contribution to others.

Although the mind itself is always in the past or future, planning, worrying, reminiscing, comparing and judging, reality itself is only ever in this very present moment. Our existence is just a whole chain of present moments, influenced by the past, building to the future, yet lived only now. Therefore, if we aren't being present, we are missing out on fully experiencing our lives and relating to those around us. How many parents interrupt telling their kids a bedtime story to answer their phone? What message does that send to the kid?[94]

How many people have conversations where they are just waiting for the other person to stop talking so they can share what they think and how they feel? How often do we do one thing, while thinking about something else entirely?

Like most other spiritual paths, Judaism is very focused on bringing the practitioners consciously into the present moment. Jewish practice emphasizes the value of stopping and becoming aware of where you are, what you are doing and why you are doing it.

There are two main parts to mindfulness – awareness and intention.

- Awareness means becoming highly conscious of what is happening in the present moment, externally and internally, without judgments or comparisons. It means being centered and present, not just going through the motions. If we were truly

---

[93] My theory is (and I'd love someone to design this research) that if someone in reality liked place A more than place B, but the photos of place B got more likes and comments, later on they will 'remember' liking place B more than place A.

[94] I used to turn off my phone when I read my kids stories in bed at night. Now I leave the phone on a bit quieter, so when it rings and my kids say "Abba, your phone is ringing,' I can say 'I know, but what's more important, the phone or you?' before carrying on reading.

aware of what we were doing, we'd never leave the house and forget if we'd locked the door or turned off the oven and we'd not let unwanted negative thoughts keep going through our minds.

In Jewish practice there are many things we say and do to promote this awareness. For example, there is a phrase that we say before each mitzvah we do which begins with;

'Behold, I am prepared and ready to perform the mitzvah of ….' This brings us into the moment and focuses us on what we are about to do. The vast majority of mitzvahs themselves are physical acts; lighting candles, shaking a lulav and etrog, tefillin, tzedakah, which give us an anchor in the present moment in the physical world. Yet they are much more than just that.

A remarkable distinction that I have found in Jewish mindfulness is the reality that not only do we need to tap into the present moment, but that there are specific present moments, in which there is a different energy or spiritual reality we need to become aware of. Our sages teach us that at different times of the day, week, month and year there are different energies being 'pumped into' creation, energies that can only be tapped into by doing specific acts. For example, the energy present in creation just before sundown on Friday night is very different to that an hour after night fall on Pesach. As the sun is setting on Friday night the present moment calls for lighting candles, on Pesach night we need to eat matzah. A Jew sitting in a temple in the Himalayas meditating and gazing at the moon on the 15th of Nissan can be very present to feelings and thoughts and sensations, can feel very blissful and connected to their souls and all of existence, yet the Torah teaches that if he isn't eating matzah, he isn't fully tapping in to the intrinsic Divine energy and nature of that moment. It doesn't mean he is a bad person - indeed, theoretically, he may be much nicer, friendlier and more spiritual than someone eating matzah in Jerusalem (thinking about their business interests). It's just that they are missing a huge opportunity to be connected to Hashem in the highest way. The Torah is a guidebook letting us know which ritual we need to perform in order to create the vessel to receive the Divine energy of the moment. The word spiritual is made up of two words - spirit and ritual. Just feeling connected but without performing the action is not enough, yet also just performing the action without any feeling is missing the point. (If I keep telling my wife I love her but don't do any actions to show that, there's something missing and if I do the actions, buy her flowers and give a massage but without any emotional investment it is better than nothing but doesn't really build the relationship).

- The second part of mindfulness is intention; being conscious of why you are doing something (your motivating factor) and what you are hoping to achieve. The above phrase we say before doing a mitzvah often continues with 'In order to give nachas to our Creator.' Although the word nachas is colloquially translated as pride or a pleasant feeling, the phrase literally means the 'landing of the spirit' i.e. bringing Divine consciousness and presence into this act. It could also be read as to settle my own spirit down and focus it towards the Creator. This phrase therefore tunes us fully into what we are doing and why we're doing it... Behold I am about to do the mitzvah of xxx in order to feed my soul and bring Gd-consciousness into my life.'

One major idea we see from this is that in Judaism mindfulness is by no means the end goal. Mindfulness of the present moment, observing and accepting what is, is important and feels great and blissful. Tapping into 'The Power of Now' is a fundamental tool we need in our lives to find peace and personal empowerment.

However it is only a means to an end, a tool to assist us in achieving the real purpose which is expressed in perhaps the key verse in the Torah:[95] 'Now Yisrael, what does Hashem ask of you, ONLY THIS, to have יראה - yirah!

Yirah, often mistranslated as fear but more accurately translated as awe, comes from the word רואה – to see. The Torah is urging us to work on becoming G-d-conscious, to find and see G-dliness in a world in which Divinity is so hidden. With a physical body, negative traits and mind patterns, pressures from life and no obvious revelation of Hashem in the world, the task of connecting in an authentic way is extremely challenging. Yet this challenge is the whole point of our time in this world, so that we work for and earn the connection when it is not so apparent. A relationship we have to work for and earn is much closer and more greatly appreciated than one we are given for free. Just being blissfully mindful, without having a relationship with Hashem, feels great but is missing the whole point.

One of the key tools to bring this awareness into everyday life is the recital of one hundred blessings a day. If said with consciousness and intention, it means one hundred mini connection points to the present moment spread throughout the day, every day! Everything we eat, every mitzvah we perform, even every time we go to the bathroom, we have a moment of mindfulness and appreciation as we say a blessing and connect to what we are doing and why we are doing it. The word Baruch can mean the source of all blessing, it can mean an increase in something and also could mean to 'lower down.' This implies that every time we say Baruch we are 'pulling down' G-d-conscious into the world, thereby standing 'face to face' with Hashem at which point we can truly address Hashem as Ata – You, showing our intimate relationship with a G-d who is imminently present in our lives.

---

[95] Devarim 10:12

Mindfulness of eating plays itself out through the laws of Kashrut, what we can and can't eat, waiting between meat and milk, saying brachot, eating special foods for different occasions. The laws of modesty help us be mindful of what images we let enter our consciousness and how we interact with people, including our spouses, in a respectful and healthy way. Mindfulness of time is expressed through Shabbat and holidays, on which we get to unplug, turn off the screens and notifications and reconnect to ourselves and our loved ones. Prayer services give us these little islands in time every day, morning, afternoon and night, helping us to realign our focus and values.[96] Mezuzas on our doors are a reminder each time we enter into another room to raise our consciousness of Hashem in the present reality. Of course, even prayer and brachot can be superficial and meaningless if ancient Hebrew words are just mumbled with no thought, just an obligation that we have to get out of the way before we can get back to things we deem more important. A person can learn Torah, pray and say blessings all day, yet completely miss the point of life if he is doing it unconsciously and without intention and joy and in fact G-d himself said that He is not interested in mitzvot done this way. [97] The Divine system is there, set up to give us constant reminders to reconnect to the present moment and tap into the purpose of life, it's our job to interact with and implement it in the right way.

The Baal Shem Tov taught that the greatest obstacle to real spiritual growth and connection is doing things habitually. He learns this from the first meeting of Hashem and Moses.[98] Hashem says to Moses של נעליך מעל רגליך ..... Take off your shoes from on your feet! The second

---

[96] According to Jewish law one shouldn't do anything before praying in the morning apart from those things that are preparing you for prayer. This means that by the time you have got up, washed your hands, got dressed and prayed it's at least an hour and a half before you can even check your phone!

[97] See Yeshayahu 1:13

[98] Shmot 3:5

half of the verse seems rather redundant; Moses knew where his shoes were! The Baal Shem teaches that the word for shoes - עליך has the root meaning to lock and the word רגליך - feet has the root הרגל - a habit. He teaches that Hashem is saying that to have an authentic relationship with Him Moses will have to unlock his habits, do things fresh and new every time. We see this idea every week in the prayer service to receive the energy of Shabbat when we sing שירו לה שיר חדש - Sing to Hashem a new song.[99] The truth is that it's not new at all! We sang it last week, and the week before and the week before that and David HaMelech wrote it 3000 years ago! The point is that we need to approach it with newness each week; that since we have grown and are new people, we see it from a new perspective every week. It's not the same me or the same moment as it was last week. When people ask me how they can do the same thing every day without it becoming stale or habitual, I ask them back if they ever get tired of telling their children how much they love them. Once you truly value something and love it, it never gets old. See Appendix G for some exercises and practices you can implement in order to integrate mindfulness into your daily life.

Putting these things into practice you'll soon find that following the Torah in a conscious and joyful way leads us towards a meaningful and mindful life, in which we can build an authentic relationship with and connection to the Ever Present One – the One who is always in the present moment. King David said 'Taste and see that Hashem is good.' What does he mean by 'taste?' You can't taste Hashem. We all know that taste is an experience. You can never tell someone what something tastes like - they have to bite into it to experience it for themselves. An experience is not in the mind, it isn't made up of thoughts. Experience is an emotional sense. We can talk about, learn about, sing about Hashem all day and yet miss the whole point which is to have an emotional experience of Gd-consciousness. Hashem is a reality that we need to

---

[99] Tehillim 96

become palpably aware of and Judaism is a system of spiritually empowered mindfulness tools leading us down the path to achieving this.

Gazing across Phewa Lake I could see the pure white pagoda perched on the top of Ananda Hill, brightly reflecting the afternoon sun, beckoning me to make the pilgrimage to meditate in its surrounding gardens. The Shanti Stupa is one of eighty such structures set up by a Japanese monk after being inspired by his meeting with Gandhi in 1931 to spread peace and love throughout the world. It had been two days since my return from my Annapurna Base camp trek and it was time for a new adventure, time to separate myself from the tourist hub that made up the one side of the lake and get over to the other side where people only tended to visit on day trips. By this time, my Nepalese was pretty good so I knew I could make a go of it. There were locals waiting on the banks of the lake offering boat rides to the famous Tal Barahi Temple on an island in the middle of the lake, or across to the stupa on the other side. After a bit of bargaining, including a couple of Nepali folk songs to lighten the mood, I persuaded one of the men to lend me his boat so I could row myself to the opposite shore where I'd leave it for him to pick up next time his cousin went across. I set off and having hit the center of the lake, decided to pull in the oars and lie back to appreciate the clear blue sky for a while and relish the freedom of being completely alone once again. As dusk started to fall I decided to make my way to the other shore, and as the day visitors were beginning to return I had the feeling of triumph once again as I rowed against the stream, off to another unknown destination and the adventures it promised.

I parked the boat on the shore and started making my way up the steep hill that lay behind the thin beach, not sure what I'd find. In a clearing half way up the hill I was greeted by Binod, a young Nepali guy with a huge smile, clearly delighted for the company and hopeful for the business. He had a ramshackle two room 'guesthouse' that had clearly seen better days. Seeing as it was already getting dark I decided to stay for the night and make my way up to the stupa the next morning. We lit a fire and cooked some dahl baht, which we ate while sitting on his roof, playing the bamboo flute and singing folk songs until the heavens were lit up by the myriad stars of the Himalayan night sky. One day turned into two, two into three and in the end Binod's little guesthouse became my home for just under two marvelous weeks. Every morning, I'd rise before the sun and make my way up to the deserted stupa which was beginning to be silhouetted against the dawn sky. I became friendly with Durega, the caretaker; a simple man from the village who always wore a bright orange tracksuit top, white shorts and sandals and who was valiantly trying to wean himself of his alcohol dependence. We'd sit inside for a while banging the large taiko drums, repeating a famous Buddhist mantra over and over again as the sun rose over the mountains, painting the sky an awesome mixture of oranges, reds and light pink, before the first tourists started to trickle in. My days were spent between the stupa and Binod's guest house where I ended up helping out, greeting and making lunch for travelers on their way up to the pagoda and even getting some of them to take the other room that I wasn't using. During the day, I'd help schlep bricks up the hill to build a small meditation cell and then go off to find a quiet spot to focus on my breath and rise beyond the confines of the physical world. One day, I had the privilege to go to a wedding in a nearby village where I met Binod's family and managed to make excuses as to why I couldn't marry his fourteen year old sister. In the early evening, I'd go back up to the stupa and help Durega tidy up after the tourists had left, have a final meditation session of the day, before making my way home where Durega, Binod and I, and occasionally a traveler who stayed the night, would cook up some dinner, crack open some beers, and chat and sing the night away.

At the time, those twelve days felt like some of the most wonderful of my life. Sitting on a patch of grass, the grandeur of the Annapurna Massif as a backdrop, clear blue skies, no sounds apart from the birds singing and the breeze rustling the leaves on the trees, no company apart from the eagles flying overhead and the ants and beetles scurrying past, just a book, a piece of fruit and some water in my backpack, no responsibilities, no-one needing my attention, able to just tap into the peace of the present moment. It was truly blissful and at the time I felt no lack.

Yet many years later, sitting alone in the Jerusalem forest, clear blue skies, no sounds apart from the birds singing and the breeze rustling the leaves on the trees, no company apart from the ants and beetles scurrying past, just a book, a piece of fruit and some water in my backpack, that present, clear, empty space that I felt in Nepal was now present, clear and *full*; saturated with an authentic personal connection to the Divine Creator of all things, as I was beginning to taste and see the goodness of Hashem.

Chapter Nine
Japan
Forty Day Pilgrimage and Aikido Training

"*Come - attack me however you like."*

This is the proposition my aikido sensei issued to me on my first day in his dojo. I already had a black belt in taekwondo and had trained in shaolin, so was a little hesitant to attack this gentle and dignified older Japanese man.

"Come on then, what are you waiting for?"

Cautiously, I moved towards him and before I knew it I was on the floor with my arm pinned behind my back, face squashed against the training mats.

"While you are here, you can attack me whenever you want; in the changing room, on the street, if you see me in a restaurant - just go for it."

There was no challenge or arrogance in his voice at all. It was very matter of fact. In fact, he said it quite lovingly, as if he were offering to treat me to dinner.

It took me less than a week to realize that it was not a good idea to take him up on his kind offer, rather, the first lesson I learnt was to humble myself literally at the feet of a true master...

Enlightened Jewish Master   Japan

In writing this book no other place has brought up such a deep nostalgia in my mind and fondness in my heart as has Japan. Nowhere was I happier and more fulfilled than I was during my fifteen month stay in Tanabe, a small town on the southern coast of Honshu, the biggest of Japan's four main islands. The strikingly diverse natural landscapes where the Shoguns and ninjas once roamed, the society still permeated with Zen Buddhist mindfulness, samurai stoicism and Shinto based spirit and animistic beliefs, the pristine beaches, colorful temples, ancient shrines and remote hiking trails through the mountains, all mixed in with modern efficiency, technology and cleanliness make Japan a truly magical place to live. Tanabe is sandwiched, (or should I say 'sushied,') between an immaculate white sand beach and mystical rolling hills which are dotted with shinto shrines, waterfalls and onsen - natural hot springs. The small town is the entranceway to the Kumano Kodo, a sacred pilgrimage trail which makes its way from one coast to the other across Japan's largest peninsula stopping at three of Japan's holiest shrines on the way. Tanabe, like most places in Japan, has several magnificent local matsuri (festivals) in which the whole town comes out to parade colorful costumes, playing traditional music and performing folk dances through the streets. It is home of the ume boshi, the world famous pickled plum whose power was said to be so great that a samurai could walk for a whole day on just one of these small delicacies. I was enamored with Japanese culture, from my apartment with its traditional matted tatami flooring and shoji - sliding paper doors, to intricate and refined traditional customs such as Chanoyu - the Tea Ceremony and Ikebana - flower arranging, both inspired by Zen mindfulness meditations. From enthralling tales of heroic samurai warriors and their martial art and survival feats, to the amazing variety of regional cuisine, Japan was truly a feast for the senses.

I had a wonderful group of local friends, the youngest of whom was about 65 years old, with whom I played tennis, went hiking, soaked in onsens and had lavish meals spiced up with warm sake - rice wine. I loved my work teaching English in an after school academy to Japanese people of all ages and the other gaijin (foreigners) in town were great to hang out with. I had Japanese lessons, sang traditional Japanese songs in the karaoke bars, trained in Aikido, played the piano in classy (and not so classy) jazz bars, and was gifted a large suzuki motorbike by a teacher who was leaving and couldn't sell it. I used to ride off into the mountains for solo hiking expeditions, scrambling over rock formations, along streams and emerging at the top to be greeted with panoramic sunset views of Tanabe and the coastline stretching out as far as the eye could see in both directions.

Tanabe is also the birth place, home and resting place of Ueshiba Morihei, founder of modern Aikido and according to some 'the greatest martial artist to ever have lived.' The dignified and honorable art of Aikido seemed like the appropriate one to study there, and my many hours of meditation in the mountains and next to the grave of O Sensei (the Great Teacher) helped me tap into pulse of the town, the culture, the history and Aikido itself - the Way of the Harmonious Spirit.

"I suddenly forgot all the martial techniques I had ever learned. The techniques of my teachers appeared completely new. Now they were vehicles for the cultivation of life, knowledge, and virtue, not devices to throw people with."   - Ueshiba Morihei

Beyond just a defense and fighting system, martial arts in the East were fused with spirituality and religious ideals, becoming a way of life and path to self-perfection. Practitioners focused on what is the appropriate action and reaction in any given situation, physical, emotional, mental and spiritual. They delved into meditation and mindfulness, discipline and focus, morality and healing, connection to the Tao[100] energy of the universe and going with the flow of life.

I'd like to say that I was on the level to embody this spirit of martial arts in my training, yet, seeing as was a little over excitable and immature (one master said to me "you know you don't need to make those silly noises…" referring to my Bruce Lee-esque shrieking), and I was getting that spirituality from meditation in nature, my martial arts was mainly about physical training, fighting styles and energy release. I completely reveled in the hours of intense training, practicing the same foot movements, blocks, stretches, kicks, punches and forms thousands of times. Although I was never very flexible, I had one main asset; my energy. Hold up pads for me and I could keep kicking them all day, spar with me you'd have to knock me out to stop me coming at you. I had to punch out candles without touching them, jump onto walls and rebound into a kick with the same foot to break wooden boards, jump through hoops and over chairs and chop through the boards with the tips of my fingers in knife hand.

Added to that was all the weight and stamina training – hours in the gym, early morning runs along the beach or up the surrounding hills and excruciating stretching routines to open up my hips for those good ol' spinning roundhouse kicks. I relished every day of it as I reflected on the fact that I was actually living my own real-life Kickboxer / Rocky 4 training scene.[101]

The one thing that really stood out in training was the immense attention to detail. My masters worked on how I was standing, breathing, moving, the angle of my fist, the way I looked at my adversary, what I could eat and even what I was thinking. To be a black belt, to avoid getting

---

[100] The Tao can be roughly thought of as the *flow of the Universe*, or as some essence or pattern behind the natural world that keeps the Universe balanced and ordered – Wikipedia.
[101] The films with arguably the best training scenes - check them out on YouTube!

knocked out, to be able to act and react in the highest way possible takes precision, attention to details and lots of practice.

This turned out to be a huge insight for me as I was starting to connect to my Jewish roots. One of my biggest challenges with Judaism was all the seemingly superfluous little details. Wash your hands like this, eat exactly this amount of bread, don't use the same plate for milk and meat, can't say this, can't do that. At the time, it seemed to me that Jews were caught up the whole time in meaningless rituals made up by a group of men who didn't have anything better to do than make life unbearably dull and difficult, whereas Buddhism was about presence, inner peace, enlightenment and other such universal and expansive ideas. I thought to myself "If G-d is the infinite loving Divine Creator of all reality, why on earth (or heaven) does He care if I tear toilet paper on Saturday? Surely He should be spending His time sorting out war and famine, not caring if I put my right shoe on first or stick a small box on the doorposts of my house."

Then one day it hit me.

In all areas of life, when we value something, details count. The sports fans who watch the video replay to see if the ball was half a centimeter over the line or not, the CEO running the business, the heart surgeon doing their work, the husband making his wife her tea just how she likes it … attention to detail gives life it's depth and meaning, it's what creates real connection and personal mastery. Far from Hashem being an angry dictator in the sky and the Torah being His book of laws for me to follow (or else!), I started to see Hashem as the ultimate martial arts master and Jewish laws as the tools guiding me towards living the most conscious, moral and intentional life possible. I realized that by mindfully following the system of halacha (Jewish law), I get to live fully and meaningfully, connecting to the present moment, raising my level of consciousness and moving through life having the appropriate thoughts, speech, actions and reaction in all situations - in other words, being a true martial artist in the way Ueshiba himself defined it. The Zohar[102] teaches that the 613 mitzvot are actually pieces of advice guiding us towards self-realization and the word Halacha comes from the root 'to walk' meaning that halacha teaches us how to walk through life consciously and intentionally.

Take putting your shoes on, for example. The law says that you must put your right shoe on first then your left. Seems a bit pointless, until you realize that it means that you have to take a second to stop and be present in the moment as you choose which shoe to put on. Then, learning a little deeper, we understand that right represents chesed – loving kindness. We want to start our day with loving kindness, bringing loving kindness into our lives and making sure we interact with others in this way throughout the day. (The bride and groom both step under the chuppah with their right foot first, symbolizing their intention to build a home based on chesed). Then, left represents Gvurah meaning strength and boundaries, as we need to rein in the unconditional love sometimes so as not to overwhelm others or be taken advantage of. If I put on my left shoe first, G-d isn't going to freak out and punish me. It just means I lost an opportunity to have a conscious connection to the present moment, thereby turning a mundane act into something meaningful and holy. This goes for all actions throughout the day – how we wash our hands, talk to people, go to bed, wake up, go to the bathroom, walk – every act during the day is a portal and a connection point to higher consciousness.

Once, as I was becoming more observant, one of my friends said to me, "You're so unlucky. You used to be so free, now you have to make a bracha before you eat anything and you have all these laws you have to follow – don't you feel so restricted?" I explained that I don't *have to* make a bracha, I *get to* make a bracha. It's an amazing opportunity to uplift eating to being a holy practice beyond just the physical act itself.

---

[102] Foundational book of Kabbalah from 2000 years ago.

Then, (speaking in the best Bruce Lee voice I could muster - try it yourself), I explained that

"Freedom is not defined as having no rules. Just eating whatever you like, smoking whatever you like, doing whatever you want does not mean you are free. It just means you are a slave to your desires, moods, ego, conditioning and what your society expects of you. True freedom comes from understanding the point of the game and mastering the laws so you can fully express yourself and move towards the goal."

Someone who wants to express themselves fully through art needs to be familiar with the laws of shading, tone, perspective and depth. A great composer needs to understand pitch, harmony and what each instrument adds to a melody, chefs need to know how much of what spice, how long it should cook for, what wines go with what food and in martial arts, if you don't know the rules and forms you're going to get knocked out. Why should spirituality and personal mastery be any different?

The word Torah comes from the root Hora'ah – meaning instructions, teaching us that the Torah is an instruction manual for living the most powerful, meaningful, conscious, moral, conscientious, joyful, connected life possible.

I used to believe that becoming more observant and following the rules meant giving up my Self for what the Rabbis and G-d told me to do and be. Yet I found that, to the contrary, it meant finally becoming free to fully express myself beyond my society, conditioning and desires; to experience the depth and joy that comes from living in line with my soul purpose.

I still have extremely far to go to becoming a Tzaddik and Mastering Life; but at least I'm very clear on the goal, how to get there and am committed to putting in the effort. This gives me a fighting chance. When I understood the path more deeply, I started performing the mitzvot more much joyously and consciously. In this way I soon saw that the Torah, mitzvot and the details of Jewish law, far from being restrictive and burdensome obligations, are in fact the Divine keys we need to help us become real Black Belts in life.

"The journey is best experienced walked solo. It's tough, but that's when the learning and insights occur." Posted on the O-Henro forum.

As soon as I heard about the 1200 kilometer circular pilgrimage around Shikoku Island to 88 Buddhist temples there was no amount of samurai on horseback, or ninjas hiding in the rooves who could stop me from giving it a go. Okay, so maybe my aikido master could have stopped me single-handedly, literally, but nonetheless I set my heart and mind on the adventure from the moment its existence was brought to my attention. Every year, thousands of O-henro-san (pilgrims) undertake to visit each temple along the route, where on arrival they perform elaborate rituals, including washing hands at the gate, ringing the bonsho bell, offering incense and coins, visiting each of the different temple buildings on site to recite specific goeika (sacred poems) and honzon mantras, practice meditation and chant the Heart Sutra – said by many to be the foundational teaching of Mahayana Buddhism. Set out in a 280-word treatise, I had memorized the Heart Sutra by the time I reached temple number 17 and would repeat it many times over as I walked the trail trying to focus myself on the path to satori – sudden enlightenment.

Shikoku Island is split up into four prefectures, which to the pilgrim represent different stages on the journey of life. The trail starts in Hosshin no Dojo[103] - the Dojo of Awakening Faith, where the yearning for and belief in something beyond takes root. From there the henro enters the Shugyo no Dojo - the Dojo of Religious Discipline where the

---

[103] The word Dojo literally means 'Place of the Way' and usually refers to a martial arts training hall.

work of transformation and spiritual cultivation takes place. The third stage is the Bodai no Dojo - the Dojo of Awakening where the enlightening effects of the effort start to be felt. Finally comes the Nehan no Dojo - the Dojo of Enlightenment where you can sit down and bask in the splendor of life. From here, the pilgrim re-enters everyday life, but by now with a whole new perspective on how to live.

Most of the henro are Japanese tourists who complete the route by bus or car, taking ten days to two weeks to visit each temple. Many aruki-henro (walking pilgrims) spread the pilgrimage out over several years, taking a week or two in the summer to walk as far as they can and then come back the next year to start up again where they left off. Then there are the few of us with more time on our hands who set off to complete the trail by foot in one go.

As soon as my contract at the English academy came to an end, I packed my bag - a couple of pairs of shorts, a couple of t-shirts (all beige or white to identify myself as a pilgrim) a small towel, a mosquito net, a rain poncho, a padded roll up mat to sleep on and sandals to change into at the end of the day to relieve my blistered feet and set off on what is still to this day the greatest adventure of my life.

Before boarding the ferry across to Shikoku, I took the two-hour motorbike ride out to Koyasan, resting place of Kobo Daishi, the great esoteric Shingon Buddhist monk who was born on Shikoku Island in 774 CE. After several years studying and attaining enlightenment in China, he returned to live and die on his island of birth, where he set up or at least visited each of the 88 temples which now form the pilgrimage route. The visit to his resting place before and after the trip is said to give henro the protection and merit they need for the arduous journey through some of the most challenging yet glorious terrain in all of Japan.

When I arrived at temple number one, Ryozenji, I bought my kongotsue (pilgrim's staff) and trail map - which was totally in Japanese. Japan has three alphabets, only two of which I could read, making map reading somewhat challenging at times. I bypassed the other traditional pilgrim's uniform consisting of Hakui, a white waist length cotton shirt which represents a henro's burial shroud and the conspicuous sugegasa - a conical-shaped hat made of straw representing the henro's coffin, mainly because I just couldn't afford them but also because I wasn't planning on dying along the way. Over the next forty days I experienced the highs and lows, joys and struggles, epiphanies and challenges of walking the henro trail. There is nothing like being on the trail on your own, everything you need on your back, totally free to go where you like, no distractions, no responsibilities except to use your core strength and wits to keep yourself alive, fed, clean and healthy. Days turned into weeks of pushing forward with a clear goal and purpose to achieve, surrounded at all times by the majestic natural masterpiece which is Shikoku Island. You soon get into the trail mind set: the morning routine, the map reading, the temple routine, the quiet awareness of your mind and surroundings, the search for the small red pilgrim figures painted on lamp posts and fences guiding the way and the 'getting-ready-for-sleep' routine. Once the blisters, and the blisters on top of those blisters hardened, my legs just begin to carry me, propelled by no more than willpower and an almost robotic commitment to just keep moving forward. I slept in parks and forests, bus stops and train stations, on beaches and benches, in tsuyado, [104] in the temples themselves which often had rooms for pilgrims and on four or five occasions I was invited to spend the night with a kind resident who couldn't bear the thought of a pilgrim, especially a gaijin[105] one, sleeping on the ground in the local park. For cleanliness, I bathed in rivers and showered by the beach, used bathrooms for wheelchair users (more space) by filling up bottles of water from the tap and pouring

---

[104] huts scattered around the island set up especially for henro to rest in.
[105] Foreigner

them over my head, and a few times I found hose pipes in public parks to wash off the grime of a full day's walk. I did laundry in sinks and rivers and would hang it up to dry between trees or lampposts while I was resting, or just drape it over my bag as I walked. I picked up food along the way in local markets and stores and was often fed and given money by people as o-settai (a donation) wanting to gain merit for helping me on my way. I was even invited to join an office annual dinner for a company who were taking the train from the station I was about to make my home for the night. After a lavish five course meal I was invited back to sleep at one of the workers' houses, a huge step up from the packet of sushi and train station bench that was going to be my food and accommodation for the night. I was usually alone on the trail but for some stretches I found a walking partner, Ken from Tokyo, Yoshi from Kyoto, another from Nara, all who were using their holiday time to notch up their pilgrim miles for that year.

The trail passed through villages, towns and cities, along beaches, through fields and forests and, seeing as the most beautiful and peaceful place to build a temple is on top of a mountain, I spent much of my time climbing the winding paths or rock stairs that lead up to the sanctuary on top. Forty days after setting out I arrived back at Ryozenji, Temple Number One. I sat for a while contemplating and integrating what I had just been through. An epic adventure into Japanese culture, nature, spirituality and history and more significantly into my Self, my limits and my mind. On the trail there were so many particulars to think about, food, shelter, map reading, medical supplies - one wrong turn, one bad decision, could add hours to the journey, one right one could greatly ease the burdens and reveal the magic the trail has to offer. Once again, walking The Way demonstrated to me the great lesson that little details really can make a big difference

Chapter Ten
Israel
Coming Home and Walking the Land

I was sitting on a hill just outside of Latrun, a thirty eight kilometer walk from Jerusalem, two nights before Pesach. We'd just set up camp, the pot was on the fire, someone had whipped out the guitar, someone else was finishing off praying the afternoon prayer.

It was the sixteenth day of our epic 'All for the Kids' fundraising hike along the Israel National Trail, which runs 1125 kilometers from Tel Dan on the Lebanese border to Eilat, nine kilometers from the southernmost tip of Israel.

I planned to complete the walk in forty days based on the story in the Torah[106] in which the spies walked the land of Israel for forty days and gave a bad report. I wanted to connect to the land, raise money for orphanages and do a tikkun (fixing) for the spies by walking the whole country for forty days and giving a good report at the end of it. The guy in the trail office said that forty days was perhaps a bit too fast a pace and that most people did it in around two months. By that time though there had already been quite a bit of coverage in Jewish press in several countries in which I mentioned the whole forty-day thing and it wouldn't look so good if it took more than that, so the challenge was laid down.

As we tucked into our dinner of spaghetti with tuna and tomato sauce, it occurred to me that my ancestors living in the land 3000 years ago most probably spent the night in this very spot, cooking their dinner, a couple

---

[106] Parshat Shlach: Bamidbar 13

of days before completing their pilgrimage into Jerusalem for Pesach. I felt that I was living the next step in the Jewish story, finally connecting to the land of my ancestors, the paths they walked, the traditions they kept and the prayers they uttered, in the best way I knew how, by walking the Way of the ancient masters. It was the next major hike on my warrior's journey, following on from the ones on Shikoku Island and in the Himalayas; just that this time the Way felt much more like home.

The story is told of a pious yet poor Jew, Yitzchak, from Krakow in Poland. One night he dreamed of a treasure buried under the royal bridge in Prague. He didn't pay any attention to it; Prague was very far away and dreams are just dreams. The next night he had the same dream, and so too the night after that. Three times in a row was enough for Yitzchak and he decided to make the journey to Prague. When he got there he was excited to see the bridge, exactly like in his dream. It was heavily guarded with soldiers patrolling at all times. He hung around for a while trying to find an opportune time to start digging. His movements (and spade) made one of the guards suspicious so he came to confront him. Yitzchak felt he had no choice, so shared his dream and offered him some of the money. The guard burst out laughing! "You crazy old man. You came to Prague based on a crazy dream?! I also had a dream like that. I dreamed that there's a poor Jew named Yitzchak in Krakow and he has a treasure buried under his oven. Do I travel to Krakow to dig it up? Of course not! I'm not crazy. Go home now and forget about your silly dreams!" Hearing this, Yitzchak wasted no time. He raced home, starting digging under his oven and sure enough found buried treasure there![107] Yitzchak realised that sometimes we have to travel far from home in order to learn and experience what we need in order to discover that the treasure we were looking for was at home all along.

I'd been living in Asia for six years and felt that it was time for a major change of scenery. My sights rested firmly on Central and South America, in particular some psychedelic shamanic training in the jungles of Mexico and a canoe adventure down the Amazon, the world's longest river. I decided to make a stopover in Israel for a few weeks to visit some friends and family I hadn't seen for a while. I'd been to Israel several times before; a couple of weddings, representing Great Britain Juniors in the Maccabi Games, a one month Israel tour with a youth movement when I was sixteen. The last time I visited was over twelve years previously when I lived in Israel for three months, working on a kibbutz near Netanya and volunteering with kids in Dimona. Having spent much of my time in India hanging out with Israelis, I had plenty of people to visit and couches to crash on during my three week stay. As I boarded the plane in Tokyo on my way to Israel, little did I know that I'd only actually make it to South America nine years later, as a Rabbi leading a group of American Jews on a social justice and personal growth trip to Peru. My path was about to take a major turn towards finding the treasure that was buried in my back garden all along.

---

[107] It is said, legend or otherwise, that the Yitzchak Synagogue in Krakow which survived the Nazis was built by Yitzchak from the treasure he discovered in his own home.

After spending my first week in Israel being spoilt by my first cousin and his family on their moshav, I decided it was time to head out and see the land. I remembered that my favorite place was Tsfat, the cool air, maze of streets, expansive views and lofty spiritual vibe made it feel to me like the Nepal of Israel.

On the way up, I stopped off to stay in a kibbutz next to the Kinneret with an Israeli friend I had met on my travels. I spent my time hanging out with the guys there, most of whom had just come back from their post-army trip around the world. We sat around all day playing guitars and bongos, chatting, eating and going to a local spring to cool off in the midday heat – throw in some rice and dahl and you could have convinced me I had landed back in India. There was one religious guy there, David, who was very interested in my travels, and on hearing that I'd spent many years studying other religions asked me why I hadn't learnt about Judaism. I replied honestly that although I had a strong Jewish identity, Judaism and the Torah seemed to me a pretty irrelevant, outdated, elitist, ritualistic, dogmatic, male chauvinistic, closed minded, fear and guilt based, unspiritual, even immoral religion, and that religious people didn't seem to me to be too spiritual, well mannered or even particularly nice. It was a decent list.

In fact, it is because of these factors that most Jews who are seeking a more spiritual path in life go to Eastern religions which seem much more compassionate, open, loving and spiritual. I read one figure that 40% of practicing Buddhists in America are actually Jewish. It doesn't surprise me, Judaism definitely needs a new PR team. David said that he really could see why I think all these things, but that maybe I just perceived Judaism like this because I had never really learnt in more depth what we actually believe and teach. That made sense. My Jewish education stopped when I was 13 and consisted mainly of learning some Bible stories and my barmitzvah portion. It's like watching Family Guy or The Simpsons. A six year old can enjoy them on a surface level, but if you are twenty six and still watching them through the eyes of a six year old, not grasping the underlying satire and irony, they just seem banal. Reading the Torah, in English, with no commentary, is a sure-fire way to completely miss the depth and wisdom contained within it. In fact, I have often found that the more weird something seems, the more deep and powerful the real understanding is.

After two weeks there it was time to make the journey up to Tsfat. David told me about a cheap youth hostel called Ascent where I could hang out, go on hikes, meet other young people and take classes in Jewish mysticism. By the time I arrived, it was after dark and as soon as I stepped off the bus I felt the calm and spiritual uplift that I still feel whenever I visit this holy city. I walked towards Ascent and as I turned on to the road I could hear some loud music in the near distance but wasn't at all prepared for what I was about to walk into. Rather than a quiet, mystical hostel, I arrived into a crazy Chassidic party, vodka flowing freely, a live band and hundreds of Jews of all types singing and dancing in circles. I was welcomed with open arms and a glass of vodka and discovered that they were celebrating Yud Tet Kislev – the 'New Year' for Chabad Chassidim and also the yahrzeit (the day of the death) of the Baal Shem Tov's main disciple and successor Dov Ber, the Great Teacher from the Town of Mezritch.

After the warm welcome and another lechaim or two, one of the chassids asked me what my Hebrew name is. 'Ber' I replied. I was named after my grandfather Bernard, who had passed away long before I was born. They looked a bit confused. "Ber isn't a real Hebrew name, your name must be Dov Ber," they exclaimed excitedly. 'Okay,' I thought, 'whatever,' – it seemed to make them happy and everyone had another lechaim.

I ended up spending a month living at Ascent in exchange for doing some writing work and helping other guests settle in. During the days, I volunteered at Livnot ULehibanot, an incredible organization specializing in a mix of hiking, volunteering and Jewish learning, where I spent my time helping to rebuild a school that was damaged by a bomb in the Lebanon War earlier that year. In my spare time I'd take classes on Judaism at Ascent, as well as a Chabad yeshiva up the hill and I connected to some young American students studying at Yeshivat Shalom Rav where I was allowed to sit in on some of their classes. These were the first real religious Jews I had come into contact with and I was pretty impressed by their integrity, warmth, intelligence and general attitude to life. I was welcomed into homes for Shabbat meals, people were always open to answering questions and I was introduced to a whole world of Jewish wisdom and practice that I had been unaware of - including putting on tefillin for the first time in my life (at the age of twenty-nine). After a while someone suggested I meet Yonah Akiva, an American chassid who learnt all day in the large shul overlooking the ancient cemetery. I made my way down through the web of alleyways and entered the study hall which was full of people wearing black coats, with long curly peyot (sidelocks) and beards, shouting at each about whatever was written in the large tomes filled with Hebrew and Aramaic writing open in front of them. I found Yonah, who greeted me warmly and offered me a seat opposite him. He shared a bit about his life growing up in a secular home in Boston, going to college in Massachusetts and volunteering in the inner city Peace Corps before coming to Israel where he lived on Kibbutz for a few months. He had a few classes in Jewish philosophy but then left to travel through Egypt, Jordan, India and Nepal for half a year after which he decided to head back to Israel. A few more classes turned into a few years of learning, during which time he moved up to Tsfat, found a community he connected to, got married, had several kids and now spends the majority of his day teaching and learning Torah. It was amazing to me that someone who had lived a 'normal' life and even traveled a bit had eventually chosen to go down this path. It was nearly Chanukah, so we decided to learn something about the upcoming holiday. Of course I knew the Chanukah story, the Maccabees, the oil in the Temple, and that we light candles and eat doughnuts, but that was about it. Yonah opened the book in front of him and started to read and translate. I don't think I've ever seen anyone so passionate and alive with what he was learning and teaching. Suddenly, Chanukah wasn't just a nice festival commemorating what happened a long time ago, it had powerful contemporary meaning and significance. I don't even really remember what he was saying but it was something about Chanukah being the pinnacle of the year, with this supernal energy coming down and the flames of the candles affecting the higher worlds and connecting us to our souls which are a piece of the Divine soul and so on and so on. I couldn't believe what I was hearing. A religious Jew talking about G-d-consciousness, mindfulness and personal development! It was a far cry from the image I had created of religious Jews walking around a bit miserably doing a bunch of outdated rituals out of fear of punishment. Chanukah finished and I went back to learn again, I believe it was a Wednesday. I sat down opposite him and he said 'I'm glad you came today, because in many ways Wednesday is the pinnacle of the week! This energy is coming down, Shabbos is beginning ....' and once again I was whisked off into a world of wonder, passion, profound insight and joyful connection to the Divine energies and purpose of life. I walked out of there into the serene mystical energy, crisp air and clear skies of Tsfat, winding my way through quaint cobbled lanes, past vibrant study halls and synagogues, small children with kippot and peyot running home from school, surrounded by beautiful views of the Kinneret and Mt Meron. I found a quiet spot on the top of a hill to meditate, gently closed my eyes, and, although I was far from convinced of anything, I felt a deep yearning to find out more about what my own tradition teaches about the nature of reality. As I settled into my practice I felt my heart opening and my soul awakening, as I started to get the feeling that maybe, just maybe, I'd stumbled upon exactly the thing I had been looking for since beginning my warrior's journey in my small apartment in Manchester University seven years earlier.

After just over a month someone suggested that I go to Aish HaTorah in Jerusalem, a yeshiva especially set up to share the basics of traditional Judaism with those who didn't grow up in observant homes. There I could learn the foundations of Jewish philosophy and practice, ask whatever questions I had and interact with Jews from all backgrounds, while getting to live in the heart of the Old City of Jerusalem. It sounded like a reasonable plan, so I packed up, made my way down to south from the City of Air to the City of Fire, [108] checked into the Heritage House,[109] and made my way to the Western Wall for the first time in twelve years. Although I didn't have a spontaneous spiritual experience like many other people do, I definitely felt the sanctity of the area, and enjoyed meditating in the sunlight, opposite the wall that had seen hundreds of thousands of prayers, tears, songs and dances throughout the generations. Someone pointed me towards Aish and I popped in for a few classes which continued to reshape the stereotype I had created of Judaism and religious Jews. After a couple of days, I was offered a place in the dorms - and that's when the red lights started to go off back home!

---

[108] There are four holy cities in Israel related to the elements. Jerusalem is fire, Tsfat air, Hevron Earth and Tiberias water.
[109] A free hostel in the Old City of Jerusalem for Jewish travellers.

Sadly, very often when a child starts to become more religious it can cause tremendous issues for the family. It can be pretty scary for parents who imagine that their kids are getting brainwashed and joining a cult, or hurtful that they seem to be going against the values they were brought up with. Parents worry that their kids are going to end up dressing strangely, being un-relatable and refusing to eat in their home, and all in all it can feel like they are losing their child. Although unfortunately this sometimes does happen, in almost all cases, this is not the case. Over the years, I have seen that almost invariably, after the initial worries and struggles, any family who had a strong bond to begin with, one in which the kids are sensible and sensitive and the parents are open and supportive, it all works itself out in the end. Once the parents see that the kids have remained basically who they were before and are not becoming cult members and that they are actually spending time working on their character traits, becoming kinder and happier and trying to the live the healthiest most meaningful life they can, things smooth out and settle down, (especially once the grandkids start arriving!). Of course, an unhealthy person who 'finds religion' just becomes an unhealthy religious person. I was always taught that the more one is becoming spiritually refined, connected and is working on improving themselves and their character traits, the less judgmental and the more open, accepting and loving they become. Therefore, when a person starts becoming more religious and begins putting down other types of Jews, especially their family, it means they haven't actually become more religious, they've just become self-righteous.

I turned up in yeshiva kind of hoping to show the Rabbis how naïve and backwards they were by sharing with them the insights I had learnt on my journey through the East. I debated with Rabbis from all backgrounds and walks of life, some who had become religious at a later age, others who grew up in religious families, all of whom seemed to be highly intelligent and well thought out. After a while I noticed that all the ideas I was sharing with them they'd either refute with very sound logic or go and get a book from the shelf and say - 'Totally! The Vilna Gaon says that here in Even Shlema ... "The Ramchal speaks about it in Derech Hashem ..."

I started to realize that the reason I was not attracted to Judaism was that I had a very superficial understanding of what Judaism actually believes.

The truth is I was pretty taken aback and humbled. These were guys with great character traits and values, profound wisdom, worldly knowledge, and who seemed to be living conscious and joyful lives. Far from brainwashing, I found it to be a very open environment, no pressure, no judgements, just a bunch of guys delving into themselves and their heritage, grappling with the purpose of life.

Some people seem to equate religion with irrationality, blind faith and neediness, as opposed to the rational, modern, scientific world view of the 21st century. And yet, here I found very rational and intelligent people tackling the question of life from a philosophical, scientific and highly spiritual point of view, which was in no way a contradiction. I carried on learning and questioning and debating and three weeks turned into three months. I formed great friendships with the other guys getting a taste of yeshiva life for the first time, friendships that I still have to this day. Instead of teasing each other and putting each other down, which seemed to me the main mode of communication amongst guys back home, we built up a strong comradery, supporting each other on our journey, learning, singing, drinking and of course playing football together. Living in the Old City was intense and magical, we went on trips to see other parts of the land and I went to my first orthodox wedding which was so moving, with such palpable holiness, that my eyes filled up with tears for the first time in many years. I experienced the festivals and Shabbat in a deeper way having learnt more about them; and started to find the joy in Jewish ritual and practice. My parents came to visit to see what I was getting myself in to and were as amazed as I was that our impression of yeshiva being a dark, morose place, old men with long beards hunched over books, no sunlight, bare rations and rats running around couldn't be farther from the truth. They joined us for some classes and my father shared with us all how impressed he was with the guys, teachers and atmosphere. Although it may not be the path my parents were hoping for me, their fears were alleviated, and they were completely supportive like they have been throughout my life. [110] I started going to some of the prayer services and having learnt about some special Rabbis with the name Dov Ber I decided to be called to the Torah with that name. For me having a Hebrew name helped me re-identify myself with who I would ideally like to be rather than be stuck in the persona I had created for myself growing up. It was a very special time, surrounded by good people, all questioning life, consciously working on ourselves, having uplifting classes and discussions and getting to experience living within the walls of the ancient city where many of the stories that we were learning about took place.

And then it was enough.

It all just seemed a bit too good to be true. Although I was far from being a naïve, vulnerable teenager who could be swept away in it all, I felt the need to get out of there for a breather, to gain a new perspective on where my journey was taking me and reconnect with what I thought and felt to be right when I was not surrounded by it all.

My sister was living in Malawi, a small country next to Mozambique in Southern Africa, where she was working as a doctor and researcher in one of the main hospitals. She'd been asking me to visit for a long time, and this seemed to be the perfect opportunity; I hadn't seen her for quite

---

[110] At one point I was talking to my mum about how great the learning and life was and she said "but you're not wearing a kippah are you?" A few months later I was expressing that I was struggling with some of the ideas and she said "but you're still wearing your kippah aren't you!?" She could clearly see a positive change in me.

a while and had never experienced Africa (outside of South Africa). So I packed my bags, bid farewell to my friends and Rabbis and headed out to realign myself and take stock of my life in The Warm Heart of Africa.

Notwithstanding our very different life styles and opinions in some areas, my sister Danielle and I have been extremely close since our late teens. We'd already travelled together in Sri Lanka, Thailand, Japan, Korea, Bali and India, so I relished spending time hanging out with her, getting a glimpse into her life and experiencing a whole new country and culture.

She arranged for me to volunteer at the Kuunika Foundation, a school for impoverished children on the outskirts of Blantyre, and each morning I'd get the local bus out to the village, comfortably back in my familiar role as the only white man, just this time I was called Azungu instead of Farang or Gaijin. I spent my days teaching English and helping with the farming, as well as writing a business plan and doing some fundraising for them. On weekends, my sister and I would travel round the country, visiting Lake Malawi, Mulanje Mountain, one of the many exquisite tea plantations, or just hanging around town going to live music events and drinking in the local backpacker hangout. I spent some time every day reading the Jewish books I'd brought along with me, including Derech Hashem by the Ramchal and Living Inspired by Rabbi Tatz. I also started to try to implement some practices - I made Kiddush every Friday night (on not kosher wine), avoided all sea food, and didn't have ice-cream after eating (non-kosher) chicken or beef. It was a very important time for me to return to balance, realign with my values and integrate where I had come from into where I seemed to be moving. There were many things pulling me in all directions – I wasn't ready to commit to all the things Judaism would demand of me and I still had dreams of visiting South America and canoeing down the Yukon. Yet, after a wonderful couple of months it started becoming clear that I had some unfinished business in Israel. My soul was being called back to the Holy Land, and so, once again I packed my bags, boarded the plane to Tel Aviv and the rest, as they say, is 'His' story.

On my way back to Israel I stopped in South Africa to visit my family there. My parents flew in from London and we got to spend some quality time together. One evening, as we were preparing supper in my auntie's house I asked her what her Dad, Bernard's Hebrew name was - after all I was named after him. Without knowing the Hebrew name I had chosen for myself, she told us that his Hebrew name was, of course, Dov Ber.

It was after about four months of learning for around fifteen hours a day in yeshiva, gradually taking on more things in a slow and healthy process, that I felt it was time for another major adventure, and my sights turned towards the Israel National Trail. Almost 150 kilometers shorter than the Shikoku Trail, it seemed very doable. It was also an excellent opportunity to really get to know the land which by that time I was already pretty sure would become my home. This time, instead of just walking, I was inspired to raise money for the orphanages I had worked with around the world as well as some organizations in Israel. Every night I would huddle over my computer in the dorm room at Aish and brainstorm the project. Websites, fundraising, logistics, routes, press releases - once my mind locks on a plan I become obsessed and make sure it happens. Everything started falling into place and the project took on a life of its own. The start date was set, the articles started appearing in Jewish press around the world and it was all systems go. My roommate Jeremy and another guy at Aish, Yonatan, decided to come on board pretty last minute, and so with the strong driving spirit of adventure, some great company and clear goals, we set off to traverse the entire land of our forefathers. I could really write a whole book about the walk. It's totally remarkable that in one small country spanning just 424 kilometers from North to South[111] there are snow-capped mountains, forests, beaches, cities and desert. I'd tell about some of the grueling treks up Mt Hermon, Mt Tabor and Mt Carmel, about the 'Trail Angels' – people who offered food and accommodation to trail hikers, the time we bumped into a group of Thai farmers celebrating Songkran – their New Year and their shock as I joined in singing their traditional songs with them. We had a joint walk along the Netanya Promenade with the Forgotten People's Fund and visited Sderot to do activities and give out teddy bears to school kids affected by the constant bombings. I'd love to describe all the locals and other hikers we met on the way, the places we slept, the weather conditions we faced, the food we ate – including on several occasions very fresh, still very warm milk straight from the goats' udders into our bottles donated by friendly Bedouin farmers. Of course I'd write about the time I woke up in the desert in the middle of the night to see my stuff strewn all around and looked up to behold the Arabian wolf who had caused the mess silhouetted against the night sky, staring me out from about 100 meters away, and how I ran out of water ten kilometers from Eilat, on my own, in scorching heat. By the time I arrived in Eilat, $18,000 had been raised for two charities in Israel, Beit HaYeled and The Forgotten People's Fund, and Kuunika in Malawi. It was a pretty hot day and once I met the others who had taken a different route we walked straight to the beach and dived into the warm waters of the Red

---

[111] The trail itself is over 1000 kms as it zigzags across the country

Sea, a welcome relief after walking over a thousand kilometers to get there. We found a campground to sleep that night and spent Shabbat with the chief Rabbi of Eilat and his family. On Sunday, we parted ways and I caught the bus back to Jerusalem. As we sped along the road, rushing effortlessly past the same kilometers which at some points each unshaded step along rocky wadis, or up jagged cliffs became a challenge, I had mixed emotions. Triumph, accomplishment and contentment on the one hand, sadness at being off the trail and emptiness as the whole project that I had invested so much time and heart in had come to an end. Yet by the time I arrived back in Jerusalem, covering in just four hours by bus what had taken me three weeks to walk in the other direction, I was already planning my next All for the Kids social justice adventure, one which would take me back to the vast subcontinent of India.

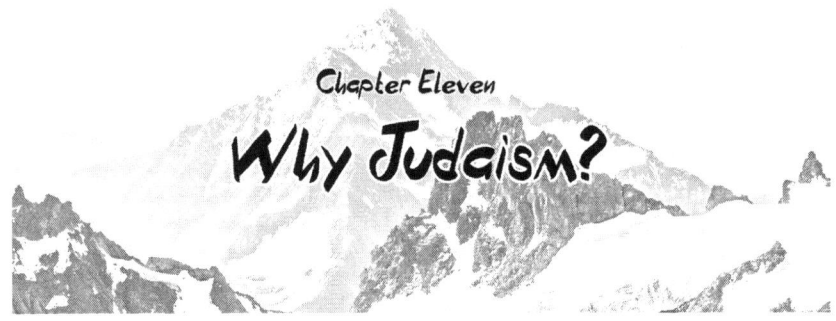

Chapter Eleven
Why Judaism?

"The challenge of this world is not finding the truth, that's pretty obvious.[112] *The challenge is having the courage to live with integrity after you know it"*
- Rabbi Leib Kelemen

The most common question that I get asked is why, after everything I saw, experienced and learnt, I ended up devoting my life to Torah Judaism. Personally, I used to think that people who become religious do so either because they are young and impressionable, susceptible to brainwashing, or lacking in happiness and meaning and therefore looking for a crutch. Yet when I landed in Israel I had no intention of even studying Judaism. I was a well-educated and experienced 29 year old not particularly susceptible to brainwashing and I was living a very happy and meaningful life.

So what did I see and learn that made me do such a drastic turn around (in a slow and healthy way)?

The truth is actually very unspiritual. In fact, it's totally rational. As mentioned in the introduction, I developed the three questions that you need to ask in order to live a well-thought out life.

---

[112] Once you've taken a serious look at the evidence.

1) What do you believe?

What do you believe is the purpose of life? Do you believe in life after death? Do you believe that you are going to hell forever if you don't believe in Jesus? Do you believe in Gd or not (once you have defined Gd)? What do you believe is the key to happiness? Do you believe not kosher food is spiritually harmful for you?

Many people have not even thought about these things. I used to ask older people if they believe in life after death. Most answers were along the lines of 'maybe' 'I hope so' 'I hope not.' It's important to take some time to think about what you believe (and then make sure you are living in line with that). There is an objective truth to existence, it would make sense to try to discover what that is and live in line with it.

2) Why do you believe that?

For most people there are two main reasons we believe what we believe. Number one is that we were told that; by parents, teachers, society. The majority of people just become a product of their society. If you were born in a village in India you'd believe in a certain Gd (depending on the village), if you were born in China you'd be communist, if you were born in Meah Shaarim in Jerusalem you'd wear a black hat. Very few people question their beliefs and values.

The second reason is that you want it to be true. It makes you feel better to think that there is life after death (life coming to an end is scary), or to think that there's not life after death (if there is life after death there are consequences). As Karl Marx said 'Religion is the opiate of the masses.' - believing in Gd makes life more meaningful and easy to cope with. We believe what we want to believe.

3) How do you know you are right?

There are so many philosophies, religions and beliefs out there, they can't all be true. Either kosher food is spiritually bad for you or it's not.

Either Muslims or Christians or Advaita Vedantans or Sam Harris or none of them have it right. They can't all be true. Either only one of them is right or none of them are. So how do you know? How do you know anything? How do you know your Mum is your Mum? Well, you look like her, she has baby photos of you, your aunties and uncles say so, you have a birth certificate, DNA test ie you have evidence.

There are three levels of belief.
 a) Faith - means I have a strong conviction not based on any evidence
 b) Belief - means I have enough evidence that it makes sense to me
 c) Knowledge - means I have beyond a reasonable doubt amount of evidence that it would be irrational not to believe.
I loved, and still do really appreciate, all the philosophies and practices I came into contact with throughout my journey. They each have insightful teachings and uplifting practices. I find that some people are hoping I'll belittle other religions, but I'd rather inspire people to connect to Judaism by sharing its depth, beauty and authenticity, than by putting down other paths. (In fact, if you need to put others down to raise yourself up there must be serious issues with you and your path). Yet this question always came up for me when studying religions and philosophies; How do you know you're right?
Of course being right doesn't make you nicer, more spiritual or a better person than anyone else. In fact, thinking they are right often makes people self-righteous, arrogant and can even lead to war. At the same time, I knew they couldn't all be right and I wanted to find out which one I should devote my life to. In all my time exploring the traditions of the world, no-one could give me a solid answer to my question - how do you know you're right? How do you know this is actually true? Just because it feels right, sounds right, someone told you it's right, I hope/want/need it to be right is not good enough. Answers such as "my master told me, I came to the insight through meditation, it seems

logical, you just have to believe" obviously didn't do it for me as there was no rational basis for the belief; he found non-self in his meditation and he found Jesus in his, his master told him one thing, his told him another. Solid intellectual belief has to be based on empirical, logical, scientific, verifiable evidence. Until this point I hadn't found any to support any of the paths I was looking into.

Until I arrived in Israel. For the first time, when I posed my question How do you know you're right? I came into contact with empirical, rational, scientific, historical evidence that the holy books of the Jewish people could not have been written without some pretty serious intervention from an existence beyond time and space, one that can make miraculous claims, knows every creature in the whole of existence and is in control of history. Although there are great spiritual truths and mystical ideas written in the holy books of all religions, there are none that a highly conscious human couldn't have come up with.

And the truly amazing thing was that the evidence was so sound that it actually overcame my confirmation bias.[113] I didn't want it to be true! I still wanted to travel, try out many different things and there were certain things in Judaism that didn't resonate with me at the time. However, after six months of grappling with it, enlisting the help of atheist forums and anyone else who would listen, I just couldn't refute the evidence.[114]

Of course, like with everything else in life (apart from the fact that you exist) there is no 100% proof, but there is definitely enough evidence that makes it unreasonable to believe that a human being wrote the Torah. After several months of debating and arguing and resisting I finally admitted that the intellectually honest thing to do would be to commit to the system that had proven itself not just to be deep, beautiful

---

[113] People find evidence for what they want to believe and ignore the evidence that goes against their beliefs.

[114] Check out Appendix I for a piece of evidence that the Torah cannot have been written by a human. More can be found at www.areyouright.org

and empowering if followed in a conscious and joyful way, but also the only one that had real evidence pointing towards its veracity. Someone once said to me that I'm crazy for believing this stuff. I thought about it and replied 'some things I believe may be a bit crazy, but I'm certainly not crazy for believing them, the evidence is all there.'

As I have shared throughout the pages of this book, I have found Torah Judaism to be an amazing system for us to get the most meaning and depth out of every area of our lives, to live our purpose, reach our potential, correct our character flaws, find authentic joy and make sure we are conscientiously reaching out to improve the world around us. Through Jewish Law (which deals with every area of life - from the moment we wake up to the moment we go to sleep), there is nothing, no matter how mundane, that can't be done in a conscious and intentional way, which makes everything we do throughout the day imbued with depth and meaning.

I have seen that all major life events, transitions and relationships are set up in a way that we can get the most out of them. For example, Aveilut, the laws of mourning, help us deal with our loss in an empowered and emotionally intelligent way. When someone passes away it's easy to either fall into depression, or the opposite, deny it and just say 'it's all for the best,' 'they've gone on to a better place,' 'they'd want me to be happy.' Judaism says that neither of these attitudes is healthy and healing. It may well be all for the best in the big picture, but that doesn't stop the pain. An integral part of the healing is to first take time to fully feel the pain, sadness, anger or guilt of the loss, hence the prescribed time period of seven days in which we sit on low chairs, people come and speak about the departed and we come very intensely face to face with our grief. Then comes a 30 day period of less intense mourning and then eleven months of saying Kaddish, plus doing mitzvahs and other positive acts in honor of the departed – all of which helps us gradually process our loss and move on, hopefully to turn our pain and loss into purpose and growth.

The laws of family purity, wherein there are certain periods that a husband and wife are not allowed to be physically intimate are not there because intimacy is frowned upon, quite the opposite. It's because of the great sanctity of marriage that the laws are specifically designed to create the most full, loving, spiritually, physically, intellectually and emotionally intimate relationship possible. In fact, many people in the secular world are starting to put this into practice as a way to keep their relationships strong, respectful and truly intimate. Added to this that the partners don't express affection for anyone else of the opposite gender. A friend once said to me that it is a bit extreme that I don't touch other women, to which I replied that it may be extreme, but that's exactly what's creating an extremely beautiful and holy relationship with my wife.

I've also learnt that Judaism provides the ultimate year round personal growth seminar. Added to the fact that there is so much Torah law about working on becoming tzaddikim, such as not to bear a grudge, not to speak negatively about people, love your fellow as yourself, and the whole mussar movement which is devoted to personal growth and perfection of character traits, I started to see that the festivals themselves are intensified pockets of time to work on particular aspects of our lives:

• We celebrate Pesach not (just) to commemorate our ancestors leaving slavery in Egypt thousands of years ago, but rather to tap into the energy of being able to leave behind our own slavery and limitations now. (The Hebrew word for Egypt is Mitzrayim – which comes from the root word Metzarim meaning boundaries/limitations). All the laws and customs are there to help us tap into this; for example we eat matzah which is dry and flat to represent humility as opposed to bread which is a symbol of 'puffed up' ego and pride. The splitting of the sea represents the fact that to be really free we need to change our nature.

• Chanukah is not (just) to commemorate that our ancestors found some oil in the temple thousands of years ago; it is to tap us into understanding that just as a candle lit in the dark can bring so much light and warmth into us, (Chanukah is at the darkest time of year), so too our greatest breakthroughs and light comes from overcoming our darkest challenges (represented by the Greeks).

In this way each festival comes to help us consciously work on a particular part of ourselves.

I also appreciate that Judaism is more practical and grounded, in line with our experience of reality than some religions. Rather than detaching from the world, Judaism encourages us to take part in it fully, getting married, having a job, enjoying good food, music, exercise and nature. Much Torah law is not about Shabbat and kashrut and lofty spiritual things which take a much higher level of consciousness to even begin to understand, but rather things such as tzedakah and business ethics, all aimed at creating a healthy society where the vulnerable are taken care of. Rather than distancing ourselves from the world, we are here to uplift and perfect it.

The lifestyle and values that I saw people living by also really attracted me. I went to my rabbis' houses for Shabbat meals and saw a whole world in which there was less focus on material gratification, I met kids who didn't spend hours in front of the TV, who were reading children's books teaching about developing good character traits and who's heroes were tzaddikim rather than men who are really good at kicking a football and then going out, getting into fights or pop stars singing songs with inappropriate lyrics.

I also used to think that to be observant I had to give up my self-expression to fit into a rigid system, yet I saw that within the structured guidelines of Jewish law there is plenty of room for self-expression. Just like everyone on the football pitch is playing to the same rules yet each player has their particular style, expression, skill set and role, so too I saw observant Jews with different clothes, different kippahs, different jobs and world views, working in offices, being doctors and lawyers and therapists, farming all day, learning all day, learning and working, going to the army, not going to the army .... I soon realized that there is no set way to be an observant Jew and I appreciated that I would not need to leave behind who I was and what I'd experienced, rather I could integrate it into a higher expression of my true self.

Certainly the orthodox world is not perfect, like all communities it has its issues. Yet I've noticed that those problems stem from human nature and exist in all communities all over the world.

So, about nine months after I first started learning in Israel, around Rosh Hashana time, I decided it was time to start taking things to the next level. I still had questions and doubts, but whenever they arose I would just go through the evidence again in my mind, reaffirming the strong foundations of my belief. Yet, it was still only another eight months later, on Shavuot that I finally committed fully to receiving the gift of the Torah and devoting my life to living with its advice. The Enlightened Jewish Master in training had finally met his true Master - and that's where the real work began.

Chapter Twelve
Soul Mates

A new ba'al tshuva,[115] I'd finally found the truth of reality,
But there was one thing missing, the other half of me,
Shteiging in the Beis, davening every day, (learning in the study hall, praying everyday)
That Hashem would bring her my way,
And then one day there she came,
Walking through the rain
And I knew that my life would never be the same ...

Chorus:
Oriana Devorah
You make me wanna learn Torah
Chassidus, halacha and gemorrah
You know I'd do anything for ya,

I wanna build a bayis ne'eman (a faithful home)
And have lots of daughters and lots of sons,
And I wanna house that is filled with shalom (peace)
And work together in Tikkun Olam (fixing the world / social justice projects)

---
[115] Returnee to Jewish observance

And I'm just so glad that our wedding day has come
And our neshamas (souls) have finally become one
So we'll sing and we'll dance and we'll have so much fun
As we give our thanks to the Ribono shel Olam (Master of the Universe)

Oriana Devorah
You make me wanna learn Torah
Chasidus, halacha and gemorrah
You know I'd do anything for ya,
And I hope I never bore ya
And you know I'll never ignore ya,
Always be there to reassure ya,
And you know I totally adore ya
Together we'll light the menorah
And you got the most beautiful aura...... Oriana Devorah

'Oriana Devorah' as performed at our wedding, 7th Tammuz 2009

Just over two years after I arrived in Israel, I met the woman who, according to Jewish tradition, was ordained for me forty days before I was even born.[116] I'd been learning full time in yeshiva for a year and a half and felt ready to take the next step. Since becoming observant I had been set up with seven different girls, each of them wonderful in their own way, but none that I felt the need to go on another date with. Then one Friday night, I was having Shabbat dinner at a friend's house, where we each shared a little about ourselves as way of introduction. One of the other guests, Nati, approached me during the meal and said that he knew the perfect girl for me. She also just recently became observant after being into Eastern religions, she is a yoga teacher, healer, social worker, artist and was born in South Africa like my parents. It sounded

---
[116] Sotah 2a

pretty promising until he told me that she had been learning in Jerusalem for a year and had just left back to South Africa literally the night before. He assured me it was a great match and he'd connect us via e-mail. I told him that as nice as it sounds, I don't want an e-mail address or long distance relationship, I want a wife and we left it at that.

I was sitting at home with a close friend after Shabbat when my phone rang. I picked up and was surprised to hear Nati's excited voice. "I thought you guys were so good for each other I called her in case I had got her date wrong and she picked up the phone!" He went on to tell me that in fact he hadn't got the date wrong at all, she *was* meant to leave on Thursday night. It turned out that she had gone to the airport and got to the front of the line where she was duly informed that the flight had been rescheduled. On inquiry when the flight would be, she was informed that it had left the night before! I've heard of flights being postponed for a few days, but never heard of one being put forward by a day. She was also a little confused as to why she wasn't informed about this; but made her way back to Jerusalem thinking to herself that obviously Hashem had some reason He wanted her to spend one more Shabbat in Israel. That thought was validated when she got a call saying there's a great guy she has to meet. It was now or never seeing as they had put her on the Sunday night flight.

So I headed out to meet "the perfect girl for me." We met at the King David Hotel and walked through a slight drizzle to a quaint café overlooking the walls of the Old City. The main difference between this girl and the seven girls I had been set up on dates with before her, was that whereas with the others it had taken me under five minutes to see that they weren't my wife, with this one I was pretty sure within about four minutes that she actually most likely was. Religious dating is particularly special seeing as once you know that you can't even touch each other unless/until you get married, the whole date is not all about 'is this going to end up on the couch at home,' and rather it's about whether we share our core values and feel we could build a loving life and family together. We chatted for a few hours and by the time I dropped her back at her seminary I was convinced my search was over. We met the next morning and decided that although it was a bit intense, she should go to postpone her flight so we'd have a few more days to date before I had to go to England on Thursday to spend time with my parents. She arrived at the El Al office and when her turn came she told them she wanted to push off her flight that night. After enquiring after her name and searching on the computer, the guy looked up and told her that it appears she wasn't even on that flight, they must have forgotten to give her a seat. Clearly Hashem was working His magic again through His favorite airline. We had another wonderful date and another one on Tuesday morning before she flew back home that evening.

I went to England to spend Chanukah with my parents and told my mum about meeting an amazing girl. I was describing her, when my mum suddenly asked expectantly "What's her name?" When she heard that her name was Oriana, my mother literally jumped off her chair and said "It's bashert! (meant to be!)" I was a little confused seeing as I was pretty sure my mother isn't a kabbalist or a prophetess. She filled me in that three months earlier she'd been in Johannesburg having lunch with her close friend from school. Her friend's daughter-in-law was there and as my mum started speaking about me and my journey the daughter-in-law told her that one of her best friends has a very similar story and was studying in Jerusalem as well. It turns out that the shidduch had already been thought of and my mum was in on it! For a few reasons they hadn't set it up, so Hashem was forced to find other ways to make sure we met.

About a week later, while I was still in London and Oriana was in Jo'burg we were chatting and she asked me how I thought it was going. She was wondering if we should still be trying to make it work, or maybe we should date other people in the meantime. I thought about it for a bit and replied quite matter-of-factly "I think you'll come back to Israel soon and we'll date and get married in June."

This, quite reasonably, freaked her out a little bit. `

However the hand of Divine Providence was on my side once again as having put the phone down she remembered that earlier that year she had been camping with a friend in the south of Israel and had had a vivid dream that she would get married the next year in, you got it, June. We dated long distance for two months, with a little breakup and make up in the middle (nothing better than a break up and make up to cement a relationship!) and eventually she arrived back in Israel at the beginning of March. A few weeks later, on our way back from an enchanting Shabbat in Tsfat, when everyone else on the bus was asleep, we were chatting quietly when I said "You know it's pretty difficult making it financially in Israel." She looked at me and smiled and said "Don't worry, I think *we'll* be okay." It seemed to really be happening and with all the pieces in place we got engaged a few days later at the beginning of April.

You may think it was pretty normal from then on in as we made wedding plans and hung out in Jerusalem, but as a forerunner to our lives, eight days later she was back in South Africa preparing to make Aliyah and I was in India where I was running the pilot program of All for the Kids' second project, Be a Kli[117]. I arrived back in Israel two months later which was just three weeks before the wedding. From the day we met until our wedding was just over six months, four months of which we were in different countries! Our holy and magical wedding was held all outdoors in Beit Guvrin on the 7th of Tammuz, 2009, which in the Gregorian calendar fell out on the 29th of, you guessed it, June.

---

[117] See next chapter

In all the ways that it's important to be similar, Oriana Devorah and I are exactly the same. In terms of our values and beliefs, our passion for Torah, Israel, healthy living and spirituality, we are totally aligned. We are both committed to growth, both personally and as a couple, we are focused on what we can give to rather than what we can get from the relationship and we use an amazing tool called 'grown up communication' when we upset each other, rather than arguing and putting each other down. Yet personality wise you probably couldn't meet two more different people. While she's quiet, graceful and calm (someone once said to me 'your wife doesn't walk, she floats!') I'm generally jumping around, wrestling with kids and singing and dancing around the house. (She once told me it's like living with Hashem the Musical). Shalom Bayit - Peace in the Home, doesn't come from people becoming the same, it comes from learning how the different energies can work together to compliment and complete each other. Rabbi Aryeh Pamensky gives an incredibly clear and empowering definition of marriage; Marriage is a commitment to putting in the effort to maintain constant emotional intimacy. That involves checking in to see how connected we are and feel, and if there is any blockage, working out and implementing whatever it takes to regain the closeness. Here's our top five Enlightened Jewish Master Couple keys to a healthy relationship whether you are looking for someone or are already married:

- The relationship must be based on sharing similar core values.
- Chesed must be the main character trait: always supporting and looking to give to the other and investing in the relationship
- Grown up communication: Never put each other down. Never speak when you are upset or angry. Find space to process then come together and share observations, feelings, needs and requests without blaming the other person.

- Growth orientated: Each person must be working on personal and spiritual growth and be committed to investing energy into the relationship.
- There must be some chemistry/physical attraction.

In our case it seems that Hashem put the Bee[118] and the Ber together to enjoy the sweet honey of life.

People often ask me what type of Jews we are, what group we identify with. The truth is that we don't fully fit into or identify as any of them. Often people who come to observance later in life don't fit exactly into any religious box. We're comfortable in all settings, learning and growing from each of them. As long as we feel someone is authentically connected to Jewish belief and practice, it doesn't matter their particular outlook on life.

No school is perfect so we send our kids to the one closest to our world view and rely on the main life education taking place, usually by example, at home. When pushed I've taken to defining myself as CJO; Consciously and Joyously Observant. I'll learn from whatever is going to help me develop myself and come closer to living with Hashem and Torah, building myself and the world around me. Our community in Jerusalem is made up of young committed Baalei Tshuva (returnees to Orthodox Judaism), most of whom have made Aliyah and are working together to raise our children with a belief in themselves, the Jewish people, Torah and a deep commitment to living a conscious and joyous way of life, strongly devoted to Jewish law, belief and practice.

---

[118] Devorah means bee

B'ezrat Hashem (with Hashem's help) we will all have the merit to raise a new generation of empowered, joyful, inspired and inspiring Enlightened Jewish Master Ninja Kids.

Our daughter Tamara Menucha was born in hospital which was a horrible experience for my wife. When she became pregnant again we decided this time we'd have a home birth. We worked with the midwife for a few months, making sure everything was healthy and going well. Around three weeks before the due date, the midwife told us that at some time in the coming week or so she would not be on call as she had a pressing personal matter to deal with. We weren't to worry, it would only be for a short time and we still had plenty of time until the due date and she gave us the number of another midwife just in case. On Wednesday the next week we welcomed a guest into our home. Rachel, a friend of a friend, was making Aliyah and we offered to let her stay in our house for a while until she found her own place. Our friend told us that she was a doula,[119] which would be an extra bonus if we needed her to help with the birth.

Just after Shabbat Oriana went into labor. We called the midwife and there was no answer. It seemed that things were progressing pretty fast so I went to get Rachel to check everything. We kept calling the midwife and the replacement one, to no avail. It looked like after all the preparation we would have to forgo the homebirth and go to hospital instead. When we suggested this to Rachel she wondered why we would do that when we could just have it at home. When I told her that with all due respect we need a midwife not a doula she looked confused. "I'm

---

[119] A doula, also known as a birth coach, is a non-medical person who stays with and assists a woman before, during, or after childbirth to provide emotional support and physical help if needed - Wikipedia

not a doula, I'm a fully qualified homebirth midwife, have delivered 250 babies and have all my equipment downstairs."

So, three days after moving into our house, we watched the amazing hand of Divine Providence work its magic once again, as Rachel came upstairs and helped deliver Binyamin Yeshayahu into our lives.

Chapter Thirteen
Back to the East

*D*iary entry from the beginning of my two month trip to India, two years after I started my Jewish journey and ten days after I got engaged.

Four and half years since last time I was here, I step out of the airport building to be hit once again by the heat, smell and noise of Mother India. Immediately swarmed by hopeful taxi drivers and beggars, I take a deep breath, relax and brace myself for what will surely prove to be another crazy adventure through possibly the most intense and multifaceted place on earth.

An hour later, I get off the still slow moving local bus in the center of town and step into the stifling, dusty heat and cacophony of car horns that is central Delhi. I start the fifteen-minute walk to Paharganj, the main market area and center of backpacker guesthouses. I used to pride myself on travelling with no more than ten kilograms, this time I have over twenty, making the walk that much tougher. The sights, sounds and particularly smells are all so familiar to me, I'm actually almost totally insensitive to it all. Is there such a thing as the opposite of culture shock?

I know how to handle the beggars, the rickshaw wallahs,[120] how to dodge the red paan[121] spit flying out of people's mouths all over the place. I am not surprised when the old man walking in front of me veers off to the edge of the pavement, crouches down in full view of everyone and releases his bowls. I nonchalantly avoid motor rickshaws, turn down offers of rides on cycle rickshaws, walk around cows, unfazed by the people lying on the side of the road, maybe dead maybe not.

None of it is new to me, I feel streetwise, an old hand at this solo travelling game. This time though, I am not searching for truth and meaning, I have already found that. I am not here on holiday, not here for pleasure, I have come with a purpose, a mission to fulfill and I am going to get it done and get back, as soon as I can, to my Kallah waiting for me in Jerusalem. I'm here to engage Israeli backpackers in social justice work with the locals, working in orphanages and teaching in schools, sharing Indian language and culture with them, as well as teaching them Jewish meditation and spirituality as a counterbalance for those seeking it in Eastern paths. As I walk down the street, the familiar sound of popular mantras is pumped out of shop fronts. Temples I once used to love going into to meditate now hold no appeal for me. I hang on to my tzitzit, and look down at the stripes on my arm left by my tefillin which I wrapped on the plane as the sun was rising across the plains of India. I find myself singing Shabbat songs over and over with a big smile on my face, drowning out the familiar words of the mantras which have come back so vividly into my head.

I arrive at the Namaskar Hotel where I stayed on my last night in India four and half years ago, not a place I would ever take my bride-to-be. I

---

[120] Wallah: a person concerned or involved with a specified thing or business.eg "ice cream wallahs"

[121] Paan is a preparation combining betel leaf with areca nut and sometimes also with tobacco widely consumed throughout South Asia, Southeast Asia and Taiwan. It is chewed for its stimulant and psychoactive effects. After chewing it is either spat out or swallowed.

go to the room and immediately fall back into my old routine - shirt off, hang up the mosquito net, check safety in the room; how the door locks, if there is any other way for someone to get in, unpack and reorganize, cold bucket shower to wash off the grime and sweat built up from the second I walked out of the airport building. I saved my kosher breakfast from the plane in case I couldn't find anything to eat. I savor the strawberry yoghurt and have a brief nap. An hour later I awake, sweating, the slow moving fan turning round monotonously, hardly providing any breeze. I blow my nose and am not surprised when black stuff from all the pollution comes out. I barely look up as a huge cockroach scurries across the room, I just hope it didn't get to the rest of my kosher food. The old essentials are still here, ten meters of string, swiss army knife, mosquito net, padlock, head torch. No huge first aid kit, no injections, no malaria tablets, just savlon (antiseptic cream) and paracetamol (painkillers) - the cures for all ills.

Yet this time, instead of my reading material being 'Autobiography of a Yogi' by Paramahansa Yogananda and 'Krishnamurti on Self-awareness,' I am carrying with me a Chumash, a Siddur, a book on reaching your potential through understanding the energy of the sephirot and the Messilat Yesharim, a classic work on personal and spiritual growth.

Now, my four cornered garment has changed from being a large woolen Himalayan poncho worn over my clothes, to my cotton Jerusalem tzitzit resting underneath my light blue shirt. Rather than listening to Tibetan mantras and flute music over and over again on a 1980's walkman which I bought in the market for $2, I have 100's of shiurim and 2 albums – Mattisyahu and Moshav Band - on my MP3 player.

Equipped very differently and with a very different goal in mind, I'm ready to face the Indian sub-continent once again...

Every year 60,000 Israelis go to India, looking for freedom, relaxation and spirituality after the army. Sometimes they'll go to a yoga class, classical Indian music lesson or go for a hike, some are seriously searching for spirituality and meaning, but in general people are just chilling out, no responsibilities, nothing to actually do or achieve. It's like being on a really beautiful spiritual kibbutz with lots of cheap restaurants and good views. I don't blame them for wanting a break - they totally deserve it. After three years in the army, being ordered around, no freedom, maybe seen a friend injured or worse, they just want to get away, treat themselves, have nothing to think about or do, no one to tell them how to think, feel, dress. Back in my early days in India I used to hang out with them a lot. Sometimes you could be excused for thinking you were in Israel; hanging out with Israelis, listening to Israeli music and eating Israeli food, just with the foothills of the Himalayas as the backdrop. They generally follow what has become known as the Humous Trail; a route of backpacker hotspots not to be missed including Bhagsu, Manali, Kasol, Rishikesh, Hampi and Goa. They were always friendly and interesting and the ones who had chosen to travel alone, away from the pack, were always amongst the most special and deep people I met on my travels. In Bhagsu where I was based, the population at times is 50% Israeli. The internet cafes all have Hebrew keyboards, the restaurants sell falafel, humous and schnitzel, there are two Chabad Houses and a Lev Yehudi House and the locals speak decent Hebrew; they have to if they want to make a living. There's a story about an Indian man who asked an Israeli 'How many Israelis are there?' to which the reply was 'Around seven million.' The Indian shook his head and said 'No, no, I meant in Israel, not in India."

Unfortunately, because a few of them could get out of control and not necessarily have the most cultural sensitivity, often guest houses would end up banning Israelis, and travelers from other countries often had their 'bad experience with Israelis' story. The main aim of my project was to give Israelis a real taste of India off the humous trail, teach them Hindi and Indian culture, make sure they were seen to be giving back, and in that way improve to some extent their reputation amongst the locals. The other aim was to engage in discussions about the Buddhist and Hindu paths they were looking into and show them the depth and beauty of Judaism in contrast.

Just like me when I was their age, western culture doesn't necessarily offer what they are looking for and Judaism, as they perceived it, couldn't be the way either. On the surface level Judaism doesn't seem deeply spiritual and enlightening, and religious Jews don't appear to be as conscious as your average Buddhist monk. Added to that, the societal problems in Israel, particularly the religious-secular divide in the country mainly to do with army enlistment, have resulted in the fact that many of my secular Israeli friends have told me that they were brought up to believe that religious people are actually mentally ill. This has pushed them to search for spirituality elsewhere, leading them further away from being open to finding out the depth and sweetness of their own heritage. I wanted to show them another side of Judaism; the meditative tradition, the social justice focus, the personal growth aspects and help them see past their, often quite understandable, prejudices. It was like a Birthright for Israelis in India, an attempt to show them that what they were searching for was actually waiting for them at home.

Continued diary entry, when I arrived in Bhagsu looking to set up the program:

I decide to walk off into the mountains to search out local schools and find out what they need and how I can help. On my first foray through the forest and around the mountains, away from the tourist area, I find a school, in Bhal Village, with just ten students. Bhal is an hour and a half trek away from the tourist area, nestled in the side of the mountain, across a spectacular river, whose blue sparkling waters rush down the valley, straight from the snow-capped mountains which form the majestic backdrop. I talk to a local boy who is manning a chai stall and learn that there are a few villages with schools dotted around the mountains. He draws me a simple map, a pencil line, a few villages, a river, scribbled names of places, approximate distances. Just my type of mission. I feel the call of the wild again, arms scratched up from making my way through thorn trees, minimal supplies, not sure what I'll find, where I'll sleep, how long before I return to civilization. All I know is that I have a mission with a defined goal and that it's taking me back, deep into the mountains, into the villages, back to the real India.

This time however, as with everything since teshuva, it is all so much fuller, clearer, so touched with the Divine. What I once called energy, nature, karma, the universe, is now something much higher; the creator of energy, nature, karma and the universe, something I can speak to and have a personal relationship with.

Saying blessings, almost constant conversation with Hashem as I make my way up and down the steep mountains and along the ridges, singing niggunim, in search of the dream, I am now part of a purposeful creation, rather than part of 'Lila' the divine sport of the Hindus, or Shunyata, the emptiness of the Buddhists. As a Jew I am here, not to transcend or disengage from this world, but rather to elevate and perfect it, thereby elevating and perfecting myself.

The stark contrast between this and where I have come from and what I am going back to literally makes me laugh out loud every time I think of it. One month ago I was studying Talmud in yeshiva, wearing a suit on Shabbat, meeting the foreign ministry to set up my India trip, planning an engagement party. Now I'm scrambling up mountain sides in rolled up trousers, sweat soaked, mud stained t-shirt and flip flops in search of chesed projects in the foothills of the Himalayas. Totally surreal, both such big parts of who I am.

Nevertheless, whereas once I dreamt of walking off and living this life, surviving in nature, living rough and simply, away from society and civilization, reaching enlightenment in some cave somewhere, or becoming a martial arts master, now the ideas don't excite me anymore. I can't wait to get married, to set up a beautiful Jewish home in Jerusalem, have guests for Shabbat, learn Torah, bring up holy children, set up organizations to help empower other people, dress respectfully to honor Shabbat and festivals. I'll still meditate, go hiking in nature and train in martial arts, yet as I look at the forty five year old travelers who have never found their place, doomed to wander around trying to impress the backpackers half their age with their travel stories, I am so grateful for having been shown my place, my base, my path, my home and having been given the person to share it with. Responsibility brings out a new depth in a person.

I have let go of many of my dreams; the three month canoe trip down the Yukon, the Amazon adventure, psychedelic shaman training in Mexico, the percussion mission to Ghana. The need to keep going in search of more adventures means you are never at peace, never settled and what you had found until now had never truly satisfied you. It did however take something much greater to be able to put those dreams aside, to outshine and replace them, something which I have found in Israel, Torah and my kallah.

I ended up running a couple of trips up into the mountains around Bhagsu where most of the backpackers never venture, giving them an authentic cultural experience off the beaten track. We schlepped all our cooking equipment and food on our back, did some pretty grueling climbs up in to the mountains, lived rough in small villages, got up early for tai chi and meditation, spent our days teaching in the schools, providing materials and painting the walls, giving Israeli music and dance sessions, and then would play music and talk meditation and philosophy deep into the starry night. There were group dynamic issues, personal issues, food issues – all of which strengthened the group, the participants and the experience as a whole. Most participants shared that it was the most powerful and meaningful experience of their trip to India, giving them a chance to interact authentically with the locals, each other, their Judaism and themselves. I too got so much out of it, being back in the hills, creating projects and bringing people together, all parts of my life coming together, it felt like my whole journey had been leading to this point. Yet at the same time my heart and thoughts were often elsewhere, with my kallah in Jerusalem and our wedding only seven short weeks away.

I really got the feeling that the time had come for this wandering Jew to finally hang up his backpack and settle down.

Or so I thought ...

In 2010, I met Rabbi Jamie Cowland who set up Justifi, an organization committed to taking Jews from all backgrounds on perspective-shifting adventures to exotic locations around the world. Based on empowering participants to take responsibility for their own well-being and that of the world around them, the trips involve meeting inspirational founders of grassroots organizations, understanding the local culture and the issues they face, doing some hands-on volunteer work and partaking in meditation and mindfulness sessions in beautiful nature. Since then, I have run dozens of trips to Thailand, Peru, Nicaragua, South Africa, India and Sri Lanka, where my whole journey began sixteen years ago.

For me, the best thing about the trip is that it brings together a group of Jews from a completely diverse range of backgrounds and Jewish affiliation to do some good and get inspired together. It's awesome to watch people interacting authentically with people they would never usually hang out with or even speak to. Each group learns to drop their prejudices, judgements and preconceived ideas, and just relate to each other as fellow Jews, usually with similar challenges on their journey through life. Often the best friendships are formed between the most opposite people.

It is essential that as a people we start to relate to each other as strongly and healthily as possible. Our strength as a people comes when we are unified, working together in mutual respect. Rabbi Moshe Zeldman asks the question: What connects us as Jews?

It's not religion, some believe some don't.

It's not race, there are many shades of Jewish people.

It's not that we're a nation, there are Jews of all different countries, languages and cultures. So really, what it comes down to is that we are a family, all Bnei Yisrael, children of Avraham and Sarah, Yitzchak and Rivka, Yaakov, Leah and Rachel. What I find so sad is that Yes, we are a family, but often we seem like a pretty dysfunctional one. Our radically different beliefs and lifestyles can push us towards distrust and confrontation, yet it is crucial that we find some way to love each other despite the differences. The greater the Jewish leader, the greater their love was for all Jews, no matter what their background and belief system. It is possible to love and respect someone even if they have different views to you. The Torah[122] teaches ואהבת לרעך כמוך - we should love each other as ourselves. That does not exclude those who believe and practice differently from us, in fact quite the opposite, it is exactly these relationships the Torah is asking us to work on. As I shared before, my Rabbis said to me that the more connected I become to Hashem, the less judgmental and more open and loving I should be becoming. If the opposite happens, I didn't become religious, I became self-righteous. The Justifi trips are an equalizer. It doesn't matter what background you are from, learning about atrocities happening around the world is shocking. It doesn't matter what your affiliation, we all need time to disconnect from the internet and work and reconnect to ourselves. It doesn't matter where you come from, playing with elephants in a sanctuary and ziplining through the forests is truly thrilling. When we can start to find common goals and common ground, working to see the good in each other, we can start to rebuild our people and go a long way to reaching our national destiny.

One morning on our trip to Sri Lanka, I was praying the morning prayer as the sun was rising over the enchanting forested city of Kandy. Suddenly, over the loudspeakers from the temples in the city below came the familiar chanting as the monks started up their morning spiritual routine. I smiled as I fondly remembered my days chanting those morning precepts, working on overcoming ego and craving. Yet, more than that, I felt grateful that my morning chants had turned from

---

[122] Vayikra 19:18

focusing on non-self and emptiness into Pesukei d'Zimra - songs of praise and prayer, directed at a relationship with the Divine presence pervading creation. From trips in India with Israeli backpackers, to Thailand and Sri Lanka with American students and young professionals, it's truly exceptional for me to be back in my old hunting grounds where everything is so familiar, to see where my path has taken me and to be reaching out to my brothers and sisters to offer them some support and insights on their own journey.

I reveled in the fact that I'd come full circle - but couldn't be in a more different place.

Chapter Fourteen
The Continuing Journey to Jewish Enlightenment

"*M*an cannot rely on intellect alone to determine his spiritual work. A connection based on intellect alone is not long lasting. He can know intellectually ..... yet his heart and body remain far behind. He needs to bind his whole soul and life force (to Hashem) and penetrate his soul to elevate and awaken it, so it becomes passionate about all mitzvot, about Torah and tefilla, and find true spiritual delight and joy in them."

A Student's Obligation – Piaseczno Rav

Fifteen years after arriving in Israel I am now blessed to be working at Aish HaTorah in Jerusalem opposite the Kotel (Western Wall), where I spend all day interacting with Jewish people of all ages and diverse backgrounds; unaffiliated, observant, people who were observant and now aren't, people who weren't observant and are moving in that direction – all searching for something more in life. It is truly heartening to see so many Jews returning to their tradition; finding depth, meaning, truth and joy through the Torah.

Rav Noach Weinberg who set up Aish, used to put the proposition as follows; Say you have a billionaire uncle and he died and left you an envelope. He wasn't your closest relative and didn't necessarily seem like the greatest person in the world, but he left you this envelope none-the-less. Would you go and collect it? Maybe there'll be nothing inside, maybe just twenty dollars, but you never know, you might hit the jackpot. Either way, it's definitely worth taking the time to check it out. Judaism is making the claim that it has access to the word of the Divine Creator, that it has the evidence to support this claim, and that it is the Guidebook towards living the best life possible. Yes, following the path does mean some sacrifices, yet if the claim is true, the effort it takes and the sacrifices I have to make are well worth it.

It's incredibly empowering and inspiring being in an environment where people are grappling for the truth, facing themselves and reality every day and striving to reach their potential. I wake up every morning feeling that I am actually living the whole purpose of life; striving to become a tzaddik forming the deepest most authentic relationship with Hashem possible, as well as guiding others towards doing the same thing.

On the other hand, it hurts me so much to see how many Jewish kids from around the world are leaving their Jewish identities behind. In fact 72% of American Jews claim to have no affiliation to Judaism at all. They know they are Jewish, but it means absolutely nothing to them. Why would it? They've never been exposed to any experience or reason that would make them value their history and identity.[123] On the other hand there are others who grew up in observant homes but something went wrong in their Jewish education, their social or family life or the lure of the secular world was just too strong. I speak to many kids who were brought up with a Judaism that was less than warm and inspiring, just being told by well-meaning teachers and parents to do things

---

[123] Anyone reading this book is invited to come to Aish and spend some time learning more about Judaism and what it means to be Jewish. Come for a day, a week, a month, a year, learn, ask questions, experience Jewish belief and practice in the holy city of Jerusalem, the Heart of the World. There are also 30 Aish branches around the world where doors are always open for Jewish men and women from all backgrounds. Visit www.aish.com Contact me directly at dbcohen@aish.com to arrange your visit

without really understanding the depth and meaning behind it. They often tell me they grew up with the idea that Hashem is an angry man in the sky and we have to do what he says otherwise He gets angry. It's hard to build a close personal relationship with a Gd like that. Jewish law can seem restrictive and formal prayer impersonal. When one walks into a Buddhist Temple in Thailand you feel an aura of peace and serenity, removed from the hustle and bustle of everyday life. Step into a Hindu Temple in India and you are met with a profound sense of the heightened spiritual energy pervading the atmosphere. Unfortunately, the same doesn't always seem to be the case when you walk into a synagogue. The Torah portion on the surface can be weird, irrelevant and even immoral and the prayer can sometimes seem dry and habitual (I wish everyone would come to our synagogue, and many others in our area - they'd see a very different picture!). Some started questioning things as they got older and came into contact with secular ideas and science, and the adults they asked were not equipped with the knowledge, the evidence or understanding to answer their questions. In fact we are facing somewhat of a crisis with a very high assimilation and intermarriage rate through no fault of anyone, just through a lack of knowledge, understanding and Jewish pride.

Looking into Judaism from the outside as I did for the first twenty-nine years of my life I wasn't too impressed either. On one hand it was because of my own. I realized that even though I had a strong Jewish identity, went to Hebrew School, had a barmitzvah, Friday night dinners, celebrated the festivals, played in a Jewish football team, visited Israel and had mostly Jewish friends, I was basically totally uneducated when it came to what the Torah actually teaches and why we believe that to be true. I had a superficial understanding of the beliefs and the rituals and was easily put off by things I saw in the Torah which went against my western secular moral value system. I had never interacted with religious Jews, happy to just judge them unfavorably from afar. In short, I was uninformed and assumed things about the religion and its adherents without really challenging myself to look a little deeper.

On the other hand, it was because of some shortcomings I saw, and am still faced with in the Orthodox community today. It's amazing how many conversations I have with highly conscious spiritually seeking Jews who come into Aish and want to know why, if the Torah is true and enlightening, there is not more emphasis on things such as physical health, protecting the environment, universalism and an appreciation of art, music and culture within the religious community. The truth is that Judaism *is* a holistic path to perfecting ourselves and the world around us. The Rambam speaks of exercise and healthy eating as prerequisites for spiritual connection[124] and the Torah and Midrash urges us to be environmentally conscious. As far as other nations are concerned a Midrash tells us that there is great wisdom amongst them[125] and that[126] that the Holy Spirit rests on man, woman, Jew and non-Jew according to their deeds. So, when asked about this I acknowledge their point and admit that religious Jews are just as human as everyone else, having their shortcomings, challenges and areas they need to really work on. Being religious doesn't automatically make someone happy and perfect; we all have personal and societal struggles to overcome and have to prioritize what areas of growth to work on. There's also some times when people who are dressed as religious Jews can behave in ways that give the religion a bad name (especially on the plane!), but they aren't representative of the religion at all, in fact it's only when they are going against Torah that the negativity comes out. However, I then feel compelled to point out that, as a whole, I have found that there is more goodwill, strength in community, welcoming of guests, giving of

---

[124] Hilchot Deot 4
[125] Eichah Rabbah 2:13
[126] Tanna D'Bei Eliyahu Rabbah 9

tzedakah, sense of purpose, focus on growth, learning and the development and teaching good values and character traits then in any other community I have ever come across anywhere in the world. The amount of charitable organizations, leadership and growth initiatives and learning opportunities is truly remarkable.

The goal is to keep our Jewish practice informed, conscious, relevant and full of vitality. The Torah and Halacha are the guidelines, leading us towards an experience. It's easy to get caught up in the laws and lose sight of the destination.

Rabbi David Aaron compares it to just looking at a menu in a restaurant without actually tasting the food, or studying the maps without going on the beautiful hike. Or, as one of my first teachers said "It is like a finger pointing away to the moon, don't concentrate on that finger or you will miss all the heavenly glory." [127]

A true Enlightened Jewish Master is one who compliments their balanced, trained and healthy mind, with a pure, good and open heart.

The most essential and powerful practice we need to bring our hearts into the picture is what the Talmud[128] calls Avodah She B'lev – the work of the heart. What is the service of the heart? It is tefilla – prayer.

The most powerful way to build a real authentic, intimate relationship with anyone is through speaking to them honestly and openly from the heart. Hashem doesn't need you to praise or thank Him. He is not needy and insecure. It is for your own good to do these things as it makes you live a life of wonder and gratitude. Hashem doesn't need you to tell Him what you want and need; it's for you to realise where everything comes from. Jewish prayer isn't an attempt to connect to a separate Being,

---

[127] Bruce Lee: Enter the Dragon
[128] Tainit 2a

thanking and praising and requesting things from Him. Rather the word for prayer - להתפלל - is reflexive ie something we are doing to ourselves! Various interpretations teach that we are evaluating ourselves, connecting to ourselves and envisioning the ultimate state of our lives and the world. To pray is the ultimate expression of our souls. You can't have a relationship with someone if you don't talk to them.

The best thing is, all you need for this is a heart and a mouth! No prayer book, no knowledge of Hebrew, no particular clothing or belief system. Just a heart and a mouth. Try out this essential Enlightened Jewish Master authentic connection exercise:

- Find three minutes a day to sit and speak to Hashem authentically from your heart. You don't need a prayer book. Just an open heart.
- Your phone should be far away from you.
- It needs to be in whatever language you feel most comfortable with to express yourself.
- It can't just be thoughts – the words have to leave your lips and you should be able to hear them.
- The more solitary a place you can find the better. Nature is good. But really you could do this anywhere[129]
- You can talk about anything – how your day was, what you are struggling with, what you're grateful for, what you are yearning for. The key is just to be real. If you feel strange doing this exercise – tell Hashem that. If you are doubting Hashem or yourself, say that. If you are upset with Hashem, let him know. If you can't think of something to say, sing a song. It doesn't matter what happens – just have a date with Hashem a couple of minutes a day.

---

[129] I once couldn't find a private place so I leaned against the wall and pretended I was talking on my phone to someone.

Judaism is less a religion than a relationship. It's a relationship between us and ourselves, us and our fellow humans and us and Hashem. Relationships take investment of time and effort in order to build emotional closeness. To the extent we learn and question, grapple and understand, explore and experience, we will develop and benefit from depth, consciousness, joy and meaning of living a well thought out life.

Chapter Fifteen
Epilogue

We're all writing and illustrating the story of our lives moment by moment and one day we'll look back and, hopefully, enjoy reading what we wrote. We just need to ask ourselves the big questions;
What are we living for? What are we yearning for? What are we investing our time, money, effort, emotion, heart and soul into? Do we fall asleep thinking about our business, our worries or about becoming tzaddikim? Are we more concerned about looking and feeling good, or being and doing good? What is going to bring us most depth, growth, meaning and joy in life?

Once we've clarified our destination and identified the path to reach it, all we need to do is put in the effort to move with passion and determination in that direction, enjoying the adventure as we go. That way we'll end our lives with no regrets, having climbed the mountains, surfed the waves, traversed the deserts, survived the falls and scaled the heights, as we walk the holy path of the Enlightened Jewish Master.

Appendix A: Chapter One
Living with Gratitude

Recognising the good:
- Choose one thing a day you have started to take for granted – your fridge, your legs, the guy who sweeps your street. Sit up straight, away from your phone, breathe deeply and take three minutes to focus on *and feel* what a great blessing it is. If you need to, imagine what life would be like without it and use the relief you feel as the catalyst for gratitude.
- Take that feeling of gratitude and direct it towards Hashem,[130] the Source of All Blessings. Articulate your gratitude in words.

- **Recognising the negative:** Notice when the voice in your head is sending you negative, ungrateful thoughts such as – 'If only I had more…' 'Not this for lunch again!' 'Why did she deserve that and I didn't.' Smile gently at those negative thoughts and say – 'I hear that, but at least I have …'

- **When expressing appreciation**, the more details the better – it makes you appreciate it more, and makes the other person feel more appreciated. Swap 'Thank you for dinner' for 'Thank you

---

[130] Hashem literally means 'The Name.' It is a general term for God but without all the baggage and false ideas that come along with that word.

for dinner, I can see how much effort you put in and the chicken was particularly good.'

- **Don't forget** to thank people for the things they are 'meant to do.' It's easy to thank people for the big things and overlook the day to day things they do for you which probably take more effort. It's easy to say thank you for a gift or a great meal, but what about the laundry and the dishes, hard work to bring in the money or collecting the kids from school?

- **Write a list:** The golden oldy. This one is very important but usually stops being meaningful as people just write something down to fulfill the task. It is only effective if you take time to not just write it down, but actually consciously feel the gratitude you have for it. To make it easier make have subtitles: My family, my health, my home, my job etc. Once in a while pull out the list and choose one to focus on or just reflect on just how many great things you have in your life.

Nice ideas? Great. Now actually make the time to do them! Don't live an 'I should have' life.

Appendix B: Chapter Two
Mental Strength and Emotional Balance

- **Basic meditation technique.** Be in a quiet place with no distractions. Phone far away. Sit comfortably on the floor or a chair – just as long as your back is straight anything goes (Lying down is not a good idea, you'll fall asleep). Set an alarm for three minutes. Breathe deeply and steadily through your nose, out through your mouth. Concentrate on your breath going in and out. If you are normal, you will be able to do this for about half a second before your mind starts wandering – thinking, planning, judging and you start having feelings – boredom, frustration and so on. As soon as you notice your mind all over the place – smile! Feel good about yourself for catching it and gently bring it back to focusing on the breath coming in and going out. Repeat the exercise. Notice that the time spent being able to focus on the breath increases and the wander time gets shorter. Notice the feeling of peace that starts to arise as you have a break from incessant thinking. Check out www.litmindfulness.org for my online meditation and mindfulness course.

- **Feeling emotions.** When we feel an emotion – loneliness, pain, sadness, anger – we often think about it and judge it and ourselves; ie Why am I lonely? It's not fair that I'm lonely. Other people aren't lonely. If I was married I wouldn't be lonely. I hate being lonely. I shouldn't complain about feeling lonely, I should be more grateful ...

Ultimately the thoughts don't help. They just perpetuate the emotion and stop you from experiencing and feeling it. Try to just breathe and feel the emotion without thinking about it. Breathe in and say "I feel really lonely/ anxious / jealous now .. and that's okay." Do this until you have really felt the feeling and see how it starts to clear up, having been experienced.

- **Write a list** of negative thoughts and beliefs, regrets and pains that often come to mind, serve no positive purpose and make you feel bad. eg 'If only ...' 'I'm so stupid...' I'll never get married.' During the day when one of them arises, rather than 'entertaining,' judging, fighting, giving in to and continuing it, just acknowledge it, smile and gently say 'Hi, thanks for coming, but you can't control me anymore.' Replace it with a positive thought.[131] Your experience of reality is dictated by which thoughts you make significant. Choose just to recognise but not to give significance to negative thoughts.

- **Equanimity/Trust Training.** Start implementing the knowledge that everything we see, hear and experience is coming from Hashem. Before getting upset, remind yourself that the guy who cut in front of your car, the missed bus, broken glasses, the bill we can't pay – they are all just challenges to help us grow.[132] Start saying Gum Zu LeTova[133] meaning 'This too is for the best,' even if you don't really feel it. After training yourself to say this all the time, you'll actually start to believe it.

---

[131] Yes, I do know that it's easier said than done. Just try it – you'll get better. Check out my class on youtube: Judaism: Mastering the mind for a full explanation of this method.
[132] That doesn't mean you have to remain passive, Hashem put them in your life to grow and then change them if you can. You can talk to the guy who cut in front of you, stand up against injustice you see, make sure you get to the bus stop earlier. It just means you skip the getting angry and annoyed bit.
[133] See Talmud: Taanit 21a

Appendix C: Chapter Three
Improving Character Traits

- **Dis-Identifying with negative character traits.**
Rather than saying "I'm an impatient person," and being resigned to that, say "I have impatience" and consciously work on it. This means you can feel good about yourself even for your negative traits, seeing as you are working on them!

- **Let it flow and let it go.** Negative character traits are manifest through thought patterns, which then create emotions and then sometimes actions. An angry thought makes us feel anger which may make us shout or hit. A powerful way to deal with the emotions is to be aware as they are arising, feel them, take a deep breath and let it pass. Don't feed the emotions by going over the thoughts that will perpetuate them ('Why me, it's not fair, how could they …') Rather, let the emotions flow, really feel them, then let them go, just like waves washing over you. Don't drown in them.

Replace the negative thought.
- Every time you notice your mind putting you down, smile and say either "That's not true" or 'thanks for pointing that out, I'll work on it."
- Every time a jealous, impatient or anxious thought arises, don't judge it, or judge yourself, don't try to escape it or make it significant in any

way. Just acknowledge it and let it go. Smile and say "You can't get me any more."

- Replace it with a positive thought ie if you have an envious thought, list three things you appreciate in your life (if you really appreciated your life you'd never be envious). If you see your mind judging someone, think of two good things about them and so on. After a while the yetzer hara stops throwing those thoughts as it sees you are transforming them to positivity.

- Make a list of the negative character traits you need to work and start consciously working on them – read books, go to seminars, seek advice and start to implement. Take small steps. Embrace the challenging situations which arise to help you work on these traits.

- Start acting like someone with a good character trait. The Sefer Hachinuch famously says

You must know that a man is acted upon according to his actions; and his heart and all his thoughts always follow after the actions that he does - whether good or bad and by doing good deeds we become good.

Therefore, if you are not so generous, force yourself to start giving more of your time and money to others. If you raise your voice often, train yourself to speak calmly. In this way you will retrain your traits.

- Finally, black belt level is building Emuna - Complete trust in Hashem. Judaism teaches that everything that we experience, no matter how bad it looks, happens for our best. If we really could integrate this, we wouldn't ever feel jealousy, anger, fear or any other negative emotion. One way to develop Emuna is to speak about it all the time. When something 'bad' happens, take a deep breath and say "Gum Zu LeTovah - Even this is for the good." You may not feel it at first, but once you know intellectually that it is true, and you consistently send that message to your system, you will start to enjoy the benefits of living with emunah.

Appendix D: Chapter Four
Self Esteem and Being a Giver

- **Self-esteem.** Write a list of ten things you like about yourself, are proud of yourself for and ways in which you help others. This means character traits and values, not physical and material things. Keep adding to the list. [134]

- Sit up straight, breathe deeply and go through the list taking time to actually feel good about these things. (Remember, just like with the gratitude list, for real depth and transformation we need to take our intellectual awareness and start to feel it in our heart. Lists don't help if we aren't 'experiencing' them).

Self Respect: Self respect comes from doing the right thing. The more you do what you know to be the good and right thing to do, you will respect yourself. How often do you say "I really shouldn't do this ..." and then do it on the spot? How can you respect yourself if you keep doing things you know you shouldn't do? How about all the things you should do .. get up earlier, exercise more, call home more ..but you don't do them?

---

[134] Watch your mind trying to sabotage this exercise by every time you think of a good trait it will find examples when you weren't perfect in that area or throw negative traits at you. Acknowledge your yetzer hara – say 'that is true and I'll work on those things but just for now I'm focusing on the positive.

Write a list of things you shouldn't do, and work on not doing them! Write a list of things you should do, and work on doing them!

- **Be a giver.** Change your attitude from 'What's in it for me' to 'What can I do for others,' from 'She's not pulling her weight in the relationship' to 'How could I pull my weight more?' (even if they aren't pulling theirs! Yeah, I know it's not fair, but Enlightened Jewish Masters aren't so worried about what is fair, they care about what is right). Do this by being aware when 'taker' thoughts arise, smile at them and replace them with 'giver' thoughts. [135]
- Give yourself a target of three acts of giving a day. That could be time, money, compliments, space, gifts. Always look for opportunities to reach out to help people.

- Make decisions based on the question 'Which of these options will help me grow most and contribute most to those around me.' (This isn't always black and white but the answer is usually the one that takes more effort. Sorry).

---

[135] Obviously you shouldn't be taken advantage of. Sometimes (although not so often - so don't use this as an justification not to give) what people need the most is for you to stop giving to them so they can learn to appreciate you and life more.

Appendix E: Chapter Five
Self Control

Try just one of these exercises a week. We're focusing on eating seeing as this is the most basic physical we have.

- **Breaking Taiva (Physical Craving):** Anything you *really* crave, ("Those fries smell amazing' 'I just need some chocolate') just don't have it.[136]

- The Rambam says you should eat whatever is good for you, whether you like it or not and refrain from things you do like if they are not good for you. With this in mind;

  - DHD! Don't have dessert. You aren't hungry and it's not good for you.[137] You only want it for the three minutes of physical pleasure it will give you. You gain much more from refraining. Start with one day a week or one time a day and build from there.[138]
  - Try eating some healthy food you don't particularly like the taste of.

---

[136] You could buy it to eat on Shabbat

[137] The Messilat Yesharim points out that no animal would choose to eat something that it knew wasn't good for it – especially when it wasn't hungry!

[138] Eventually you could build up and become a member of D.O.O.S – Dessert Only On Shabbat (and festivals/special occasions). This has the huge added bonus of increasing the pleasure and honor of Shabbat

- At a restaurant, choose what you want most, then order something else. (Or let someone else order for you)
- Take one plate of food and that's it. No refills.
- Leave some food over. Make sure it's a nice bit that you really want.

• **Building Kedusha:**
- When eating something you really like, relate the pleasure of the taste back to its source. Feel wonder and gratitude towards Hashem.
- Take the craving for the food and redirect it towards yearning for Hashem. In this way the desire itself is transformed and uplifted into desiring Hashem!

Appendix F: Chapter Six — Discipline

- **Get up like a lion.** The first law in the Shulchan Aruch, main book of Jewish daily living in the Shulchan Aruch is
Be strong as a lion (to beat your yetzer hara) to get up in the morning.[139]
Getting up like a mentsch is a complete game changer. It means you beat your yetzer hara in the first battle of the day, setting the right tone for the rest of the battles to come. It is the key to the rest of the day. You feel good about yourself, you achieve more. The Kotzke Rebbe said that it's impossible to get up like a lion if you go to bed like a donkey so:
- Go to bed envisioning getting up strongly in the morning, not dreading the sound of the alarm.[140] The ultimate level is that we should be getting up before sunrise. Every Buddhist monk, Hindu priest and really anyone who is living purposefully is getting up before sunrise. Try it once a month or once a week. The early morning hours are very powerful for meditation, exercise, goal setting. Don't sleep your life away.

- **Exercise:** The Rambam tells us the importance of exercise in Hilchot Deot

---

[139] Shulchan Aruch 1:1
[140] Don't look at your phone for at least 45 minutes before sleeping

"*maintaining a healthy and sound body is among the ways of God - for one cannot understand or have any knowledge of the Creator, if he is ill*" [141] so

"*The rule is that he should engage his body and exert himself every morning until he heats up his body.*" [142] Choose a short exercise to do – ten push-ups a day, a brisk walk to work, basic stretching routine. Start strengthening and enlivening your body - it gives you much more energy than a cup of coffee.

- **Make a list** of things you should do more of and things you should do less of. Any time you have the thought 'I really should for example
  - Get up earlier
  - Call home more often
  - Tell my husband/wife more often how much I appreciate him/her
  - Exercise more
  - Drink less coffee
  - Less time on social media
  - Find out how so and so is doing ….

  Just do it! Pick up the phone, set the alarm, close the computer, whatever it takes. If you should do it, then do![143] You may need an 'accountability partner' – someone who holds you to your goals.

---

[141] 4:1
[142] 4:2
[143] For more tips on this watch my class on youtube called When Willpower is not Enough

Appendix 6: Chapter Seven
Mindfulness

Try these Enlightened Jewish Master Mindfulness Practices

- **Make time to be alone.** Most people do whatever they can to not be alone - no phone, no book, no newspaper. We don't like being alone because our minds run wild and we need to distract ourselves. Rav Itamar Schwartz teaches that the power to be alone is deep within us and is a vital need for the soul. Adam was created alone, the world was silent when the Torah was given, many of the forefathers were shepherds so they could spend time alone. Take five minutes to just sit and Be. Get used to spending time with yourself without running away. You'll find that you're actually pretty great company.

- **When doing an activity, try to be present through your senses.**
  For example when washing the dishes feel the warm water on your hands, listen to the water splashing on the dishes, look at the droplets on the plates, smell the dishwashing liquid. be aware of your emotions, let go of thoughts about the past and future.

- **Before you say or do something** ask yourself – Would my best self do this? Why am I doing this? Note down every time you succeed.

- **Whenever you arrive in a new place**, take time to just look, observe, listen, smell, notice how you are feeling – take it all in. You can do this anywhere – just stop for a few moments and observe everything.

- **Think to yourself** that this could be the last time you get to do or see this thing.[144] When you walk past a tree on your way to work, take time to stop and look at it – start finding wonder in simple things.

- **Limit your time on the phone.** Set certain hours, or a certain amount of times a day you can check. Go for walks in nature or meet your friends for a coffee without having your phone there.

- **List some bad habits** (slouching/biting nails/compulsively checking phone). Once you have listed them you'll become more aware of them. When you notice yourself doing them, smile and correct them.

- **Say a blessing before you eat something.** Before eating take a moment to have gratitude, recognise Hashem as the source of all your blessings and contemplate how many stages and people it took to get the food to you. Articulate these things in words. At first you can use your own formulation, then you can learn about the set spiritual wording given to us by our Sages (There are many resources on aish.com about the what, why and how of making blessings).

---

[144] This is based on Chovot HaLevavot Shaar Cheshbon HaNefesh. Don't try this consciousness tool if it will make you afraid or morbid.

Appendix H: Chapter Eight
Personal Accounting

Every Enlightened Jewish Master knows that the key to real success and sustained growth is cheshbon hanefesh - making an accounting of our actions. There is no way of becoming a tzaddik if we are not charting and keeping tabs on our life. A business person has a spreadsheet for every area of the business; finances, employees, social media, marketing, goals etc because they don't want to miss an opportunity to make money or trip up and lose any money. So too, if we are truly committed to growth, we need to set goals and evaluate our progress, eliminating bad habits and retraining ourselves towards greatness. The Messilat Yesharim teaches that there is

"a need for a person to be meticulous and weigh their ways each and every day like the great merchants who continuously evaluate all of their business matters in order that they not degenerate. They should fix definite times and hours for this weighing so that it not be haphazard but rather with the greatest regularity for it yields great results."

The truth is that life is quite simple. There are just three steps:
1. Work out what the goal is
2. Understand what steps must be taken to reach the goal
3. Do them!

With this in mind, an essential Enlightened Jewish Master Training Exercise is to make yourself an Excel Spreadsheet (or do this in a notepad) called Becoming a Tzaddik/Enlightened Jewish Master.

- In the top row you'll have column headings of areas to work on to achieve the goal.
eg: Health, Finances, Relationships, Spirituality, Charity …
On the one hand, the more comprehensive we can be, the better chance we have of succeeding. Yet we must also prioritize certain things at certain times and always take small steps, making sure we don't fall into the trap of taking on too much thereby losing everything.
- Down the side you'll have:
Daily, Weekly, Monthly, Yearly.
Set yourself some SMART tasks for each of these areas. SMART stands for Specific, Measurable, Attainable, Relevant and Timebound. Vague, broad and overly challenging goals will never reap rewards and will soon make you give up. You don't have to fill in all areas all at once. Take small steps and succeed rather than large leaps and fail.
- Make sure you open the spreadsheet every day and at very least, even if you haven't achieved any of your tasks, check off the 'Charted' box to sho w that at least you opened the chart. That in itself is a huge success.
- It's best to have a mentor or accountability partner/growth chavruta to make sure you are moving ahead steadily. Give them access to your spreadsheet.

Appendix 1: Who Wrote the Torah?

There are several things written in the Torah that no human would have written or could have known. One example comes from the laws about kosher food. I can understand the Hindu / Buddhist stance of a fully vegetarian diet - we don't want to kill animals. However, limiting some animals and sea creatures doesn't make sense. Why would a human making a religion limit the food they can eat? It's a bad move practically and also doesn't make sense either in terms of survival or in terms of trade and economy. Every other nation living on the Mediterranean coast trades in sea food and has it as a major part of their diet. What possible benefits could there be for having these laws?

Even crazier is that the Torah teaches that for a mammal to be kosher it needs two special signs - chew the cud and have split hooves. It then goes on to say that there are four animals in the world that only have one sign and not the other, and that they are not kosher. Firstly, why do you have to tell me that? I know they aren't kosher; you just said you need two signs to be kosher. Secondly, and more importantly, how could a human being ever make that claim thousands of years ago living in the Middle East? How did they know every animal in Australia, China, Alaska, South America? All you'd need to do is find one more animal that has one sign not the other and that would prove the book false.

There is no benefit to saying it and it makes your book easily falsifiable. Incredibly, no fifth animal has been found.[145]

The same goes for seafood. The Torah says that to be kosher it needs to have fins and scales - like salmon and tuna. Just fins - ie sharks and whales, or neither fins and scales - ie lobster and shrimp are not kosher. The Talmud claims that there are no sea animals that have scales but no fins. Once again - How could a human know everything under the oceans? Yet true to its word, no sea creature has ever been found with scales but no fins.

Added to that is the fact that the Torah says that the nation can not do any farming for a year! How can an agricultural society survive with no farming!? Especially when their animal and seafood supply is limited. The author of the Torah promises a triple crop in the sixth year to make up for it. How can a human promise a supernatural triple crop? What happens when the sixth year comes and there is no triple crop? The book would have been proven false immediately.

The Historical Revelation

The Torah, believed by Orthodox Jews to have been given to Moses about 3,300 years ago, contains a number of predictions that are highly unusual, inconsistent and/or supernatural.

More astounding is the fact that every single one of the unusual, inconsistent and/or supernatural predictions contained in the Torah has, over the course of history, <u>actually transpired!</u>

According to biblical critics, the Five Books of Moses were compiled by humans more recently – around 2,500 years ago. Imagine the possibility of human authors, sitting around thousands of years ago,

---

[145] This is a very over-simplified version without reference to arguments and counterarguments, just to give you a taste of some evidence.

thinking what to write in their guidebook. "Don't kill", "don't steal" - sounds good to me. However, when one reads the Torah in some depth, one finds that, apart from these basic precepts, it contains some pretty remarkable claims.

Could a human, or group of humans, living many centuries ago, have written a book, containing unusual, inconsistent and/or supernatural predictions, which actually came true? Why did they even put predictions in their book at all?

Consider the following absolutely unique, supernatural history of the Jewish people, which was all predicted in the Torah and, then decide for yourself whether this could possibly have been predicted and written by humans.

Eternal

Close to the beginning of the Torah, the author claims that the Jewish people will be an eternal nation.

" I will ratify My covenant between Me and you and between your offspring after you, throughout their generations, as an <u>everlasting covenant,</u> to be a God to you, and to your offspring after you." (Genesis 17:7).[146]

Anyone familiar with world history knows that this is something that any nation would want to claim, but that is pretty unlikely to be manifest - given the fact that nations and empires rise and fall all the time. There are no Aztec, Babylonian or Edomite nations around today – to name a few. No other tribes mentioned in the Torah - Canaanites, Chitites, Jebusites - are around today.

---

[146] All translations of the Torah quoted in this paper are from the Stone Edition of the ArtScroll Chumash, 2016 impression.

Of course there are many nations which have been around for as long as the Jews, just look at China and India. However, a Chinese text written thousands of years ago saying they would be an eternal nation doesn't exist and would be unremarkable if it did. As they have a strong, large, well-established country, many many people, a shared language and culture and being generally free from persecution, you would expect that.

Yet the author of the Torah makes three predictions that go in the face of this first one actually coming true. Far from predicting that the Jewish people would have a common well-established land, many people and resources and remain free from persecution - the author predicts the exact opposite!

First, why would anyone predict such awful things for their nation? Secondly, seeing that they did all happen, how did this first prediction - that there would always be Jews - also stand the test of history?

As Mark Twain famously said:

"If the statistics are right, the Jews constitute but one quarter of one percent of the human race. It suggests a nebulous puff of star dust lost in the blaze of the Milky Way... The Jew has made a marvelous fight in this world in all ages; and has done it with his hands tied behind him. He could be vain of himself and be excused for it. The Egyptians, the Babylonians and the Persians rose, filled the planet with sound and splendor, then faded to dream-stuff and passed away; the Greeks and Romans followed and made a vast noise, and they were gone; other people have sprung up and held their torch high for a time but it burned out, and they sit in twilight now, and have vanished. The Jew saw them all, survived them all, and is now what he always was, exhibiting no decadence, no infirmities of age, no weakening of his parts, no slowing of his energies, no dulling of his alert but aggressive mind. All things

are mortal but the Jews; all other forces pass, but he remains. What is the secret of his immortality?"
(Essay "Concerning the Jews", 1899)

This does not make Jewish history unusual, it makes it completely unique.

Let's look into the Torah text itself to see these predictions.
PART ONE – Three negative predictions

1. Exiled and Dispersed

The Torah says that we will enter the land of Israel but then, after not following the Torah

"the Lord will scatter you among all the nations, from one end of the earth to the other…" (Deuteronomy 28:64)

Far from having a strong consistent homeland, the Jewish people were exiled twice from their land - first, when Solomon's Temple was destroyed in 586 BCE and second, after 'Herod's' Temple was destroyed in 70 CE.

An earlier example of Jewish dispersal is the conquest of the northern Kingdom of Israel by the Assyrians in 722 BCE, when the ten tribes of the northern Kingdom were exiled from the land and became the "Ten Lost Tribes". We see throughout history that the natural consequence of being conquered is assimilation into the conquering power's culture. This applies even more so after exile from the land completely.

For centuries, Jews were expelled from many countries eg England (1290), Spain (1492), Portugal (1496) and never had a stable homeland.

We became the 'Wandering Jews', dispersed all over the world. The vast majority of Jews today do not live in the same country in which their grandparents were born! For much of our history, most Jews never set foot in Israel. We also see historically that exile of a whole people hasn't been that common. Most empires just conquer and rule, not exile. Why would whoever wrote the Torah predict this highly unlikely occurrence?

The striking fact is that the prediction of dispersal came true. It is perplexing to think that human authors would make this claim: Would they have said: "We're going to go into the promised land - and then mess it up and get kicked out!?" What other religion makes claims against themselves like that?

We see that this prediction is:
a) negative (goes against human psychology to write such a thing);
b) extraordinary (why would humans think this would happen?); and,
c) inconsistent (goes against the eternal nation claim)

2. Few in Number

Now, suppose we were exiled and dispersed, but there were billions of us; then. no doubt, we would easily have survived. However, the Torah says:

"You will be left few in number, instead of being like the stars of the heavens in abundance, for you will not have hearkened to the voice of HASHEM, your God." (Deuteronomy 28:62)

This is remarkable considering that one of the first commandments in the Torah is to be fruitful and multiply. An ancient people with high birth-rate means a huge population. We know that, whereas the average birth-rate in America is 1.7 children per family, the average number of children in orthodox Jewish families is 6.7. According to natural history, had the Jewish nation grown at the same rate as the rest of the world, we would now be 400 million people, the third biggest nation on the planet after China and India! Far from that, we make up only 0.2% of the world's population.

Once again, who would make this prediction about their own nation, how did it transpire contrary to all natural history and how did we survive it?

It may be said it is because of persecution that we've remained few in number. However, since the Holocaust (after which we have not been overly persecuted), the rest of the world's population has grown by three and a half times. That means we should have gone from 11 million Jews at the end of the Holocaust to 39 million now. We know that isn't the case and, far from even doubling our numbers, we have maintained a consistent rate of increase throughout the ages to reach 14.7 million today (even given our overall high birth rate).

We see that this prediction is:
    a) negative (goes against human psychology to write such a thing);
    b) extraordinary (why would humans think this would happen?); and,
    c) inconsistent (goes against the eternal nation claim)

3. Persecuted

Can we think of any way an exiled nation with very few people could actually survive even though they are scattered around the world? How about this: The Jews will be so beloved in the eyes of the entire world that the nations will support them and make sure they continue to exist, no matter where we are. We see the reverse is true.

Anti-Semitism is the most universal, irrational, extreme type of hatred the world has ever known. This was, of course, predicted in the Torah:

"And among those nations, you will not be tranquil, there will be no rest for the sole of your foot; there HASHEM will give you a trembling heart, longing of eyes and suffering of soul. Your life will hang in the balance, and you will be frightened, night and day, and you will not be sure of your livelihood." (Deuteronomy 28:65-66)

David Lloyd George said
"Of all the bigotries that savage the human temper there is none so stupid as the anti-semitic. It has no basis in reason, it is not rooted in faith, it aspires to no ideal, it is just one of the dank and unwholesome weeds that grow in the morass of racial hatred."

Once again, who would write this about themselves and how are there any Jews left, given this reality?

We see that this prediction is:

a) negative (goes against human psychology to write such a thing);
b) extraordinary (why would humans think this would happen?); and,
c) inconsistent (goes against the eternal nation claim)

The author of the book has set themselves up for a lose lose situation.
Either the prophecies come true which would mean there is no natural chance of survival, or they don't come true which means the book is false.
Read that sentence again.

This might be why Blaise Pascal when asked by King Louis 14th of France for evidence of the supernatural replies - "Les Juifs" - The Jews. No other nation exists that has been exiled, dispersed, few in number and consistently persecuted for 2000 years. Exile equals extinction. The

Jewish people on the other hand, not only survived, but actually became 'the most formidable people who ever walked the earth." (Winston Churchill).

PART TWO – Three positive predictions

So far, we have seen three negative and extraordinary predictions which are not things people would normally predict about themselves. These are inconsistent with the prediction that we will be an eternal nation, and yet they all actually occurred and against the odds we survived. .
Now, let's say we could find a reason that somehow this small nation has survived (eg, there was always a core who held deeply to their beliefs), what sort of influence and impact would you expect the small nation to have on the affairs of mankind?
At best, the Jews should be hobbling through existence, tails between their legs, living in some small villages somewhere, hanging on to their traditions as best they can. Somewhat like the gypsies or aboriginals, who are pretty scattered and disliked, without a homeland of their own – and, as would be expected, have made absolutely no impact on civilisation whatsoever.
Yet, with the Jewish people - we see the opposite is true!
Consider the following three amazing predictions in the Torah of the flourishing of the Jewish people and the Land of Israel, which are being revealed in front of our very own eyes.

4. **Light unto the Nations**

The author of the Torah makes the far-fetched prediction, going in the face of all natural possibility - that this exiled, hated, tiny nation, far from being unheard of and struggling to hang on, would in fact be a light unto the nations!

"And I will make you into a great nation; I will bless you, and make your name great, and you shall be a blessing. I will bless those who bless you, and him who curses you I will curse; and all the families of the earth shall bless themselves by you." (Genesis 12: 2-3)

Incidentally, the phrase 'Light unto the Nations' comes from the Book of Isaiah:

"I the LORD have called unto you in righteousness, and have taken hold of your hand, and submitted you as the people's covenant, as a light unto the nations" (Isaiah 42:6).

Now, everyone knows that Israel, despite being just 75 years old, having no natural resources and having been at constant war or threat of war, is a world leader in high-tec, medicine, science, agriculture, defense, and higher education; and that Jews, despite making up only 0.2% of the world population, have won 22% of Nobel Prizes. That is simply not natural.

Let's read more from Mark Twain:

"Properly, the Jew ought hardly to be heard of, but he is heard of, has always been heard of. He is as prominent on the planet as any other people, and his importance is extravagantly out of proportion to the smallness of his bulk. His contributions to the world's list of great names in literature, science, art, music, finance, medicine and abstruse learning are also very out of proportion to the weakness of his numbers."

Much more importantly, it is Judaism that gave the world the morals which we all hold as obvious today, and taught the world about the existence of the One G-d in whom the majority of the world's population now believes Plato and Aristotle believed in infanticide and the Romans forced gladiators to fight to the death and had other blood sports as popular entertainment., Judaism, through the Torah, demonstrates to the world that that was not in order.

Or as John Adams, Second President of America said:

"I will insist the Hebrews have [contributed] more to civilize men than any other nation. If I was an atheist and believed in blind eternal fate, I should still believe that fate had ordained the Jews to be the most essential instrument for civilizing the nations ...
They are the most glorious nation that ever inhabited this Earth. The Romans and their empire were but a bubble in comparison to the Jews. They have given religion to three-quarters of the globe and have influenced the affairs of mankind more and more happily than any other nation, ancient or modern."

An exiled, small, severely persecuted people, becoming a Light unto the Nations?! Unheard of and totally unpredictable by humans at least 2,500 years ago! Yet, amazingly, this actually happened, as predicted in the Torah.

5.   **Interdependency of the Land of Israel and the Jewish People**
This is where it becomes truly astonishing . We know from the 1st century historian Josephus that the land of Israel was an extremely lush, fertile area, with big cities and an abundance of trees and crops. It was part of what was known as the Fertile Crescent, where the soil was particularly good for agriculture. Yet, whoever wrote the Torah made a supernatural prediction about the future of the Land:

"I will make the land desolate; and your foes who dwell upon it will be desolate. And you, I will scatter you among the nations, I will unsheathe

the sword after you; your land will be desolate, and your cities will be a ruin." All the years of its desolation, it [the land] will rest...". (Leviticus 26: 35)

The Torah is saying that when the Jewish people are exiled, the land itself will rest and remain desolate. Does the land have consciousness? Does it know who is there and refuse to bear fruit for or be settled by anyone apart from the Jewish people? Who's ever heard of such a thing? When the native Americans were driven from their land, did it remain desolate? No, it turned into Manhattan.

The fact that the land, of what is now Israel, remained desolate is even more remarkable given the fact that it is the land-bridge between Africa and Asia, therefore making it one of the most prized pieces of real estate in the world. Every empire who ever tried to rule the world came and conquered and tried to make the land flourish and habitable - yet not one of them ever succeeded. We see historically that the area of the Land of Israel did indeed remain desolate for 2,000 years - through the Roman, Byzantine and Ottoman Empires. There were no flourishing fields, no big conurbations, no thriving civilisation and a dwindling population (in 1890, there were only 440,000 people in the whole land, whereas today, Jerusalem alone has twice that number) - until 1948 that is, when the Jews came back to the land. Added to this is the fact that the conquering nations where either Christian or Muslim, meaning they certainly held the land in high esteem and would want to civilize it.

Now, there are bustling cities, commerce, technology, agriculture and a multi-million dollar export industry of fruit, flowers and vegetables around the world. History has shown that since the Jewish people arrived in the Land of Israel, the land itself has become fertile. It has responded only to the Jewish people. No other nation has ever been able to cultivate it and build a thriving civilization there.

Fertile and thriving 2000 years ago when Jews controlled the land, fertile and thriving since 1948 when the Jews came back, with a 2000 year gap of desolation in between. That is unique and supernatural as it transcends the laws of nature.

Yet, whoever wrote the Torah predicted exactly that! Was that the work of human authors at least 2,500 years ago?

6. **Return to the Land**

To top it all: whoever wrote the Torah made the incredible prediction that the exiled Jewish people, unlike any other nation ever, would not only survive, but will actually return to their land!

"Then HASHEM, your God, will bring back your captivity and have mercy upon you, and He will return and gather you in from all the peoples, to which HASHEM, your God, had scattered you." (Deuteronomy 30:3)

HASHEM, your God, will bring you to the Land that your forefathers possessed, and you shall possess it; He will do good to you and make you more numerous than your forefathers." (Deuteronomy 30:5)

A nation, or a people, who were without a homeland for 2,000 years and which still exists is unique in itself. For that nation then to return to their land after all that time is totally unimaginable. For that fact to be predicted 3.300 years ago is supernatural. Other nations have been conquered, but their populations were not expelled in the way the Jews were from their homeland.

In 1900, only 0.5% of Jews were living in the Land of Israel. In 1948 and 1949, 341,000 Jews made Aliyah. Within the first six years of the Jewish state being re-declared in 1948, the population of Israel doubled. I made Aliyah in 2008 along with 13,701 other Jews. In 2020 around

49% of the Jews of the world live in Israel. By numbers, more Jews live in Israel today than at any other time in history.

120 years ago, no-one in the world spoke Hebrew. Today it is a revived language spoken by millions of people around the world.

Jewish history is unique and supernatural and yet it was all clearly and accurately predicted. The Torah contains many things that a human wouldn't have written and couldn't have known or guaranteed.

Given these facts, it seems to me irrational to claim that this is the work of human beings in the Middle East at least 2500 years ago.

For more check out www.areyouright.org

About the Author. Follow and Stay in Touch

After spending six years living in Asia, delving deeply into Eastern philosophies, partaking in several silent meditation retreats, extreme martial arts training, hiking in the Himalayas and 1200 kms around a Japanese island, as well as spending time volunteering in orphanages, Dov Ber moved to Israel and ended up discovering the depth, beauty, truth and wisdom of his own religion. After making Aliyah in 2008 and alongside learning in yeshiva, he set up a social justice organisation, All for the Kids, which raised money and awareness for orphanages in India, Africa and Israel and made it into the Israeli Ministry of Foreign Affairs' web-booklet entitled 'Young Leaders of Israel 2010.' He's now a senior Lecturer at Aish HaTorah World Centre in Jerusalem (aish.com) and Founder/Director of Living in Tune: Authentic Jewish Mindfulness which runs on-line Jewish Mindfulness Courses and in-person retreats.

Learn more from him:
www.rabbidovber.org
Online Mindfulness Course: www.litmindfulness.org
Spotify - Living In Tune Dov Ber Cohen
Instagram / Tik Tok: RabbiDovBerOfficial
Subscribe to his YouTube Channels:
@masteringlifeseries and @livingintune
Be in touch directly - dovbercohen@gmail.com

Made in the USA
Middletown, DE
29 January 2026

27737174R00121